MY CLIMATE ACTION

Sustainable Living through
Self-Driven Actions

Dr. Manjunatha L.R, PhD

notionpress.com

INDIA · SINGAPORE · MALAYSIA

ISBN

Paperback: 979-8-89777-686-3
Hardcase: 979-8-89961-256-5

Index

Foreword *5*

Preface *7*

Chapter 1: Snow in the Desert, How and Why? 11

Chapter 2: The Industrial Revolutions and Climate Change 17

Chapter 3: From Bullock Carts to Fuel-Guzzling Cars 27

Chapter 4: The Plastic and Concrete Revolution 33

Chapter 5: Global Climate Agreements – Promises and Progress 45

Chapter 6: Heavy Industries – Challenges and Opportunities 53

Chapter 7: The Sustainable Development Goals (SDGs) – A Global Blueprint 75

Chapter 8: E-Mobility – Driving the Future 95

Chapter 9: The Energy Transition – Breaking Free from Fossil Fuels 113

Chapter 10: Net Zero – A Personal Vision 127

Conclusion 147

About the Author 149

Foreword

It is a pleasure to present *My Climate Action— Sustainable Living through Self-Driven Actions by* Dr. Manjunatha L.R, a valued colleague with a deep commitment to sustainability.

Climate change is one of the most urgent issues before us today, and it calls for responsibility at every level. While policy and industry have a major part to play, this book reminds us that individual action matters just as much.

Dr. Manjunatha brings this idea forward in a practical and thoughtful manner, showing how everyday choices can support a more sustainable future. He speaks about reducing consumption, using better materials, embracing renewable energy, and moving towards greener ways of living.

What I value in this book is its strong belief that awareness must lead to action. It encourages readers to look at their own choices with honesty and purpose.

I believe this book will be useful for individuals, professionals, and organisations who want to contribute in a meaningful way. I congratulate Dr. Manjunatha L.R. on this important work and wish it the wide readership it deserves.

Parth Jindal
Managing Director
JSW Cement & JSW Paints

Preface

How Beautiful This World Was. Or Maybe, How Beautiful It Still Is.

I remember one summer holiday many years ago. I was ten years old. It was a Sunday, and I had nothing to do. Everyone at home was busy with their own work. I sat near the window, looking outside at the bright afternoon.

Then I thought, "I will go and get an ice cream." Back then, we didn't spend much time at home staring at phones. If we wanted something, we went outside. I put on my sandals and stepped out.

The sun was shining, and the air felt warm as I walked. The street was lined with big trees on both sides. Their long branches stretched out over the road, making patterns with their shadows. The breeze was soft and felt cool on my face.

The road was empty. There were no cars or motorbikes. A boy on a bicycle passed me, ringing his bell. That sound was the only thing I could hear.

As I walked further, I saw a few women sitting outside a small house. They were making garlands with jasmine flowers. Their hands moved carefully as they tied the flowers together. They were talking and laughing while they worked. I could smell the sweet scent of the flowers as I walked past them.

A little ahead, the road turned near a field full of marigolds. The flowers were bright and orange, glowing under the sun. A soft breeze moved through them, and the flowers seemed to move together. I stood there for a moment and looked. Everything felt good.

Then I felt something wet on my arm. I looked up. A drop of water had fallen from the sky. Another drop followed, and then more. The air grew cooler, and the sky was turning grey.

The rain started suddenly. It wasn't light rain. It came down quickly and covered everything. I tried to cover my head with my hands, but it didn't help. My shirt and hair were soaked in no time.

I started laughing. The rain felt cold, but I didn't mind. The trees above me moved as the water fell from their leaves. The ground started to fill with puddles. I stepped into them as I walked, making water splash all around.

When I reached the ice cream cart, I was dripping wet. It was a small wooden cart with a colourful umbrella on top. The man selling the ice cream looked at me and smiled.

"You came all the way for ice cream in this rain?" he asked, smiling as he handed me the popsicle.

I nodded and pointed to the orange one. It was my favourite. I gave him the coins I had in my pocket.

The popsicle felt cold in my hand, while the rain kept falling around me. I took a bite. It was sweet and a little tangy, the way I liked it. Eating ice cream in the rain made me feel happy. I stood there for a while, watching the rain fall as I finished it.

But as soon as I stepped inside my house, my grandmother noticed me. *"Where did you go?"* she said. *"You are completely wet!"*

"You ate ice cream in the rain," she said. *"Let me ask you something. When you fall sick, what do we do?"*

"Take medicines," I said.

She nodded. *"Yes, but do you know why you fall sick?"*

She pointed to my chest. *"Your body is like a little world of its own. It needs care. If you eat good food, get enough sleep, and drink clean water, it stays strong. But if you eat ice cream in the rain or eat too much junk, it starts to break down. It won't stay healthy if you don't look after it."*

She looked at me and said, *"The world is like us. It needs care to stay strong. If we take care of it, the world will take care of us."*

I thought about the cool breeze on the way to the ice cream cart, the rain that smelled so fresh, and the marigold flowers moving in the field.

She said, *"If we look after what keeps us alive, the world will always give us what we need. It is our home. Wouldn't you want to protect it?"*

I had never looked at the world this way before.

Chapter 1:

Snow in the Desert, How and Why?

The desert was supposed to be hot. That's all I had ever heard about Saudi Arabia, endless sand under a burning sun.

I boarded the plane, ready for it. My bag was packed with light clothes, sunglasses, and a hat. As the plane took off, I looked out of the window, imagining the heat I would feel the moment I stepped outside.

The flight was calm. I ate the meal served and kept watching the map on the screen. It showed our plane flying closer to Riyadh. I thought about the stories of travellers who said the desert heat could knock you off your feet. I was curious to experience it myself.

When the plane landed, I stepped outside, expecting to feel a wave of heat. But the air felt different. It wasn't as warm as I thought it would be. The sun was shining, but it felt soft.

On the drive from the airport, I saw stretches of sand on both sides of the road. The sky was clear, and everything looked still. I thought, "This is just the beginning. Tomorrow, the real desert will show itself."

That night, I went to bed thinking about the heat. I was ready for the scorching sun to greet me in the morning. But what greeted me was something else entirely.

I woke up early, ready to see the desert. I pulled back the curtains, expecting to see sand and the sun. But what I saw was white. Snow was everywhere, on the ground, on the trees, and even on the roads. I stared for a moment, not believing it. This was a desert. How could there be snow?

I went outside. The air was cold. I could see my breath in front of me. I had only packed light clothes, so I started to shiver. The ground felt strange under my feet as I walked. The snow made a soft crunching sound.

People around me were surprised too. Some were playing in the snow, throwing snowballs, and laughing. I even saw a group of children trying to build a snowman. It looked more like a festival than a normal day in the desert.

Later, I found out that this was the first time it had snowed in that part of Saudi Arabia. It was a rare event, and everyone was talking about it.

I stood there, feeling cold and confused. This was not what I expected. "What is happening to the world?" I thought to myself.

I was not ready for this cold. I only packed light clothes like shirts, trousers, and a hat for the sun. There were no warm jackets or thick shoes in my bag. My clothes were too thin, and they did not help at all.

I tried to look for something warm, but there was nothing to find. The shops here do not sell heavy jackets or winter clothes because snow is not normal in this place. I crossed my arms to keep myself warm, but the cold wind still reached me. My hands felt stiff, and my ears were hurting from the cold.

The whole situation felt strange. I had come here expecting heat. I thought I would feel hot and sweaty, but instead, I was freezing. My fingers felt heavy, and my nose was running. The cold made me tired and weak, even though I had not done much.

I had prepared myself for the hot sun of the desert, but here I was, standing in the middle of the snow, shivering. It felt unreal.

When I wanted to know what happened exactly, I started asking around and reading about it. That's when I found out how rare this snowfall was. People in the Al-Jawf region had never seen snow before. This was the first time it had ever happened in that part of Saudi Arabia. Everyone was talking about it, and I wanted to understand why.

I learned that the snowfall was caused by a unique weather event. A low-pressure system had formed over the Arabian Sea. This system pulled in moist air and carried it across the region. Normally, the desert air is too hot and dry for snow. But this time, the temperatures had dropped much lower than usual. When the moist air moved over the cold desert, it turned into snow.

This kind of weather is very unusual for the Arabian Peninsula. The deserts there are known for extreme heat, not freezing temperatures. The snow covered the sand dunes and roads, creating a sight that felt almost impossible. People were shocked because they had never imagined seeing snow in a place like this.

As I kept reading, I found that the snowfall was linked to larger changes in the world's climate. Scientists have been warning that weather patterns are shifting. Events like this are becoming more common in many parts of the world. Deserts are getting colder at times, and places that used to be cold are experiencing heatwaves.

This snowfall happened at the same time as the COP29 climate conference. World leaders had gathered to discuss the impact of climate change and what could be done to slow it down. The snow in Saudi Arabia was a clear sign that things were changing.

The more I thought about it, the stranger it seemed. A desert covered in snow, people building snowmen, and children throwing snowballs, it felt so different from what I expected.

It made me stop and think about one question: **What is happening to our planet?**

What Is Climate Change

This made me think more about the world and how much it is changing. I had heard people talk about climate change for years, but I never really stopped to understand what it meant. I began to realise that everything is connected, the air we breathe, the water we drink, and the land we live on. When one thing changes, it affects everything else.

Climate change means the Earth is shifting in ways it never did before. It is not about the weather changing for a day or two. It is about long-term changes. Temperatures are rising, seasons are no longer coming as expected, and storms are becoming stronger. The planet is reacting to how we have been treating it for so long.

I started reading to understand why this is happening. I found out that the way we live is causing most of these changes. Factories release gases into the air, cars and planes burn fuel, and forests are being cleared. All of this adds carbon dioxide and other gases to the air. These gases trap heat around the planet, making it warmer. Scientists call this the greenhouse effect.

This warming is leading to big problems. Ice at the poles is melting, and this is causing sea levels to rise. Many places near the coast are at risk of flooding. Warmer oceans are causing stronger storms and hurricanes. In some areas, crops are failing because the seasons are no longer reliable. Some places get too much rain, while others are drying up.

I began to see that climate change is not only about nature. It is about people. It changes how we grow food, where we live, and even how we stay healthy. It touches everyone in some way.

As I thought about all of this, it felt clear that the Earth was trying to tell us something.

The real question is whether we are ready to listen.

Chapter 2:

The Industrial Revolutions and Climate Change

A few months ago, a friend called me with an idea for a holiday. "Let's go stay in a treehouse," he said. "There's this place deep in the jungle where they've built these amazing treehouses. It's peaceful, and you'll love it."

The idea sounded exciting. I had never stayed in a treehouse before, and the thought of being surrounded by trees and nature felt perfect. We made plans, booked the place, and a week later, we were on our way.

The journey was long, but the drive through the forest was beautiful. The trees were tall and green. The air had a clean, fresh smell. Birds were flying in the sky, and the leaves made a gentle sound as the wind passed through them. It felt calm and far away from the busy life of the city.

When we arrived, the treehouse looked incredible. It was built high up, with wooden steps leading to a small balcony. From there, you could see the tops of trees stretching as far as the eye could see. It felt like something out of a storybook.

As we settled in, I couldn't help but look around and think about how it was made. The wood used to build the treehouse must have come from trees, maybe even from the same forest where it now stood. To create this place, parts of the jungle had been cleared, and trees had been taken down.

The thought stayed with me as I walked through the jungle the next morning. The beauty of the forest was all around, trees that had stood for years, providing homes for animals and birds, were now replaced with something made for humans. The treehouse was wonderful, but it came at a cost.

It made me think about how we use nature to create things that make us happy. Whether it's paper from trees or wood for a house, we take from the environment to improve our lives. There's always something gained, but also something lost.

How It All Began

As I thought about how much we take from nature, I wanted to understand when all of this started. That's when I came across the story of the first industrial revolution.

It began in the late 1700s, a time when life was very different. Most people lived in small villages and worked on farms. Everything was done by hand or with simple tools. Then came a big change. Machines were invented that could do the work faster and in larger amounts.

One of the most important inventions was the steam engine. It used coal to create power. With this, factories could run machines day and night. Trains and ships powered by steam could carry goods and people across long distances. Everything became faster, bigger, and more connected.

Coal played a huge role in this revolution. It was dug out of the ground and burned to produce energy. This energy made factories run and powered trains and ships. But burning coal also sent smoke and gases into the air. At that time, no one thought much about what this might mean for the planet. They only saw the progress, the factories, the jobs, and the growing cities.

The Industrial Revolution changed the way people lived. Villages grew into towns, and towns turned into cities. Goods that were once made by hand were now made in factories. Life became more modern, but it came with a cost that no one understood at the time.

After the steam-powered era, a new source of energy changed the world: electricity. This period, known as the Second Industrial Revolution, took place from the late 19th to early 20th centuries. Electricity transformed industries and daily life in once unimaginable ways.

Electricity didn't appear all at once. It started as an idea, something that scientists and inventors experimented with for years before it became a part of everyday life.

In the early 1800s, scientists found out that electricity could flow through wires. Michael Faraday, a scientist from England, made a big discovery in 1831. He showed how spinning a magnet near a coil of wire could create electricity. This idea, called electromagnetic induction, is still how we produce electricity today.

Then came inventions that made electricity useful. One of the most important was the light bulb. Before this, people used candles or gas lamps for light. In 1879, Thomas Edison made a light bulb that could stay on for hours. It was safer, brighter, and worked better than anything people had before.

Once the light bulb was invented, the next challenge was how to deliver electricity to homes and businesses. In 1882, Edison built the first power station in New York City. It could provide electricity to a small area nearby. This was the beginning of cities using electricity in everyday life.

Other inventors didn't stop there. Nikola Tesla, an engineer, found a way to send electricity over long distances. He worked on alternating current, or AC, which made it possible to power entire cities. In 1893, people saw this in action for the first time at the Chicago World's Fair. The fairgrounds were lit up with electric lights, and it amazed everyone who saw it.

This invention allowed factories to operate beyond daylight hours, increasing production.

Factories were some of the first places to start using electricity. They replaced steam engines with electric motors, which were faster and safer. Without the need for big boilers or burning coal inside factories, the work environment improved a lot.

Electricity also made its way into homes. Electric lights replaced candles, and people began using fans and other small appliances. Later, inventions like washing machines and toasters made life at home easier.

Each of these discoveries was built on the one before it. What started with small experiments grew into something that powered industries, cities, and homes. Electricity changed how people worked, lived, and built the world around them. It was a revolution, but it touched every part of life.

The next big shift came with computers. It all started in the mid-1900s when the first computers were built. These machines were enormous, often filling entire rooms. They could only do simple calculations and tasks, but they were much faster than humans.

In the 1970s, things began to change. Scientists developed smaller and more powerful computers. Microprocessors, which are tiny chips, were invented. These chips allowed computers to process information without taking up much space. For the first time, computers became small enough and affordable enough for businesses to use them regularly.

At first, companies used computers for basic tasks like storing data, creating documents, and organising files. But soon, computers became central to almost every part of work. Factories used them to track production, while banks used them to handle accounts. Offices no longer needed shelves of paper files because computers could store everything digitally.

In factories, automation began to take over. Machines powered by computers started doing repetitive tasks like assembling products, packaging goods, or sorting items. These machines were faster and more accurate than humans, and they could work without breaks. This made factories more efficient, but it also meant fewer jobs were needed for some types of work.

By the 1980s and 1990s, computers were everywhere. Schools, offices, and even homes started using them. The internet also began to grow during this time, connecting people and businesses across the world. Information that once took weeks to send could now be shared instantly.

From running small shops to big factories, computers are now at the heart of everything. It was a new way of thinking, where speed, organisation, and precision mattered more than ever before.

After computers transformed industries, a new wave of progress reshaped the world. Cars became more common, changing how people travelled and how goods were moved.

When I think about how we travel and live today, it all seems so normal. Cars take us everywhere, and buildings surround us in every city. But these changes did not happen overnight. They came with progress and a cost.

The automobile industry grew quickly. Cars, trucks, and buses made travel faster and easier. They also changed how we moved goods from one place to another. Most vehicles run on petrol or diesel, which are fossil fuels. Burning these fuels produces carbon dioxide, a gas that harms the environment. Transport today causes about 16% of all global emissions. Road vehicles like cars and lorries are the biggest part of this.

Efforts to make cars better started a long time ago. Engineers worked on engines that used less fuel. Electric cars appeared in the late 20th century but were expensive and not very practical. In recent years, electric cars have become more common. They offer a cleaner way to travel, though there is still a long way to go before they fully replace petrol and diesel vehicles.

At the same time, construction was booming. Cities grew bigger, and we needed more buildings, bridges, and roads. Materials like cement and steel became essential. Cement production releases a lot of carbon dioxide during its process. It makes up about 8% of global emissions. Steel production is even higher, causing 7% of emissions. Factories burn coal to produce steel, which adds to the problem.

Buildings themselves also use a lot of energy. Homes need electricity for lights, heating, and cooking. Offices and shops need power for equipment, cooling, and heating. Residential buildings cause about 11% of global emissions, and commercial buildings add another 6%. Most of this energy comes from burning fossil fuels, which adds to the challenge.

The Fourth Industrial Revolution made life easier in many ways. Cars brought us closer, and buildings gave us space to work and live. But this progress depends on energy and resources that come at a cost to the planet.

It is a balance we are still trying to figure out today.

As I learned more about how the world is changing, I realised we are entering a new phase. This is a time when new ideas like artificial intelligence, robots, and green energy are shaping how we live and work. These ideas feel like they belong to the future, but they are already becoming part of our lives today.

AI helps us understand things better. It looks at large amounts of information and shows patterns that we might not see on our own. For example, it can help predict when a flood might happen or track changes in forests to stop deforestation. It gives people the tools to act faster and make smarter decisions.

Robots are being used to make work easier. They can do tasks in factories that take a lot of time or are dangerous for people. On farms, robots plant and harvest crops, making farming more efficient and saving water. These machines help us get more work done with fewer resources.

Green technologies are helping us move away from fossil fuels. Solar panels and wind turbines are becoming more common, providing cleaner energy. AI helps us understand how to use this energy better by predicting when we will need it most. This makes renewable energy more reliable and useful.

These ideas give us hope, but they come with challenges. For example, AI and robots use a lot of energy to work. If that energy comes from burning fossil fuels, it adds to the problem we are trying to fix. This is why it is so important to make sure we use clean energy wherever we can.

This new phase of progress focuses on smarter solutions. It reminds us that growth and care for the planet can happen together when we make thoughtful choices.

Where We Are Now

It feels strange when I think about it. Nature should save us, not the other way around. For as long as we have existed, nature has been the one giving us everything, food, water, air, and shelter. It has always been there, silently keeping us alive.

Now, we find ourselves trying to save nature instead. Forests are being cut down, rivers are drying up, and the air we breathe is filled with pollution. The very things we depend on are under threat, and we are the ones who caused it.

It does not feel right. Nature should not need us to save it. It should be strong and endless, taking care of everything, like it always has. But we have pushed it too far, and now it is struggling.

The thought stays with me. We should not be in this position, but here we are. If we want a future where nature can go back to saving us, we have to make changes.

It is a simple truth, but not an easy one.

Chapter 3:

From Bullock Carts to Fuel-Guzzling Cars

It was a cold December morning, and I woke up later than I should have. Mondays are always hard, and this one wasn't any different. I had an important meeting at the office, so I rushed to get ready, grabbed my car keys, and headed out.

I'm from Bangalore, a city known for its lovely weather, delicious dosa, and endless traffic. That morning, the traffic seemed worse than usual. My car just stood there, not moving even an inch. I looked at the long line of vehicles ahead and behind me. It felt like the whole city had decided to use the same road as me.

I looked at the time and felt that little panic rising. The meeting wasn't going to wait for me, and I was stuck here, helpless. All around me were cars, bikes, buses, more vehicles than I could count. I suddenly noticed something I hadn't before, there were so many of us on the road, all trying to get somewhere, all part of this endless jam.

I had a slight cold that morning. I didn't want to use the air conditioner, so I rolled down the window for some fresh air. Instead of relief, I was hit with the smell of smoke and fumes from the vehicles around me. It made me cough, and I pulled the window back up.

It hit me then, I was one of them. My car was adding to the same emissions I was complaining about. It felt strange to realise that in this huge mess of vehicles, I wasn't just stuck in it. I was part of it.

So, there I was, late, frustrated, surrounded by honking cars, with nothing to do but sit and wait. Just another Monday morning in Bangalore.

I couldn't stop thinking about how we got here. It made me wonder how people used to travel long before cars, trucks, and planes filled our roads and skies. It felt like a world so far away from what I was experiencing at that moment.

In the past, people relied on bullock carts to travel and transport goods. These carts, made of wood and pulled by oxen, moved slowly but steadily. They were a common sight in villages and towns, especially in India and other rural parts of the world. A bullock cart could carry heavy loads like grain, firewood, and even people over long distances. It required no fuel, only the strength of the animals. The pace was slow, but it did no harm to the environment.

Bullock carts were used for centuries. They didn't release smoke or pollute the air. The only sound was the creak of the wooden wheels and the soft clinking of the bells tied around the oxen's necks. Journeys were long, and people planned their lives around the time it took to travel. For many, this was simply the way life was.

Then came the first big change, the invention of the steam engine in the 18th century. This gave rise to trains and steam-powered ships, which could carry people and goods faster than ever before. By the early 20th century, cars began to replace bullock carts. Henry Ford's Model T, launched in 1908, was one of the first cars made affordable for ordinary people. This was followed by buses and trucks, which made it easier to move goods and people over long distances.

Planes came into the picture shortly after. The Wright brothers' first flight in 1903 opened the door to air travel, which would later shrink the world, connecting cities and countries in hours instead of days. By the mid-20th century, planes had become a crucial part of global transport, moving people and goods across continents.

While these advancements made life faster and more convenient, they came at a cost. Cars, trucks, and planes rely heavily on fossil fuels like petrol and diesel. Burning these fuels releases carbon dioxide, a major greenhouse gas that contributes to global warming. Today, transportation is responsible for around 16% of global greenhouse gas emissions, with road vehicles like cars and lorries being the largest contributors.

I started to realise how much transportation has changed and the price we are paying for it. Cars, trucks, buses, and planes have brought us speed and convenience, but they have also created serious environmental problems.

Transportation is one of the biggest contributors to greenhouse gas emissions, making up around 16% of global emissions. Road transport accounts for the majority of this, with vehicles like cars, lorries, and buses responsible for about 72% of transport-related emissions. Planes and ships add to the problem, especially with international travel and shipping growing every year.

The problem lies in the fuel. Most vehicles rely on fossil fuels like petrol and diesel. When these are burned, they release carbon dioxide and other harmful gases into the air. These gases trap heat in the Earth's atmosphere, causing global warming. The demand for fuel also drives more oil drilling, leading to deforestation and habitat destruction in many parts of the world.

Over time, our dependency on fossil fuels for transport has grown. It powers almost every vehicle on the road, in the air, and on the seas. This reliance not only harms the environment but also makes us vulnerable to rising fuel prices and supply shortages.

Efforts to solve this problem have been growing. Electric vehicles, for example, are becoming more popular. They run on batteries instead of petrol, and when charged using clean energy like solar or wind, they produce no emissions. Governments in many countries are offering subsidies to make electric cars more affordable, and more charging stations are being built to support their use.

Public transport is another solution. Many cities are expanding bus and train networks to reduce the number of cars on the road. Some places are even introducing electric buses and trains, which are cleaner and more energy-efficient.

There is also hope in alternative fuels like hydrogen. Hydrogen-powered vehicles emit only water as a byproduct, making them a cleaner option. Research is underway to make this technology more affordable and practical for widespread use.

Despite these advances, the transition is not easy. The infrastructure for clean transport is still developing, and many people continue to rely on petrol and diesel vehicles. It requires a lot of investment and commitment to shift to cleaner options.

The choices we make now will shape the future of how we move and live on this planet.

It is strange how far we have come. We moved from bullock carts to cars, trucks, and planes, all in the name of progress. Life became faster, distances became shorter, and the world felt smaller. It sounds like a good thing on paper, but when you stop and think about it, you start seeing the cracks.

Modern transport has made our lives easy, but it has left the air heavy with smoke. Cars sit bumper to bumper on roads, each one adding a little more to the pollution. Planes fly across continents, leaving trails of emissions high above us. Trucks carry goods everywhere, running on fuel that keeps us moving but also damages the planet.

And now, we talk about electric cars as the solution. They seem like the answer to our problems, clean and shiny. But there's more to it than what we see. Making electric car batteries requires metals like lithium, cobalt, and nickel. Mining these takes a lot out of the Earth, digging deep, cutting through land, and leaving scars that may never heal. It is like fixing one problem while creating another.

We are always searching for something new to solve the mess we are in. We focus on technology, believing it will save us, but sometimes the answers are simpler. Look back at how people lived not too long ago. They travelled slower but lived lighter on the planet. A bullock cart never polluted the air, and walking never needed a drop of fuel.

This does not mean we should go back to bullock carts and forget about progress. But maybe we need to slow down and rethink how much of this "progress" we really need. *Do we need cars for every little trip? Could cities have better public transport to reduce the number of vehicles on the road? Could we walk or cycle more often, for ourselves and for the air we breathe?*

Somewhere along the way, we started seeing convenience as the most important thing. The truth is, we made choices that worked for us but not for the Earth. Now, we are trying to save nature when it should be the other way around.

Progress gave us speed, but it also gave us pollution. Maybe it is time to ask if this rush was worth it and, more importantly, what we can do to change the path we are on.

Chapter 4:

The Plastic and Concrete Revolution

It was the holiday season, and I wanted a break from the usual routine. My friends and I decided on a beach vacation. We didn't think twice about where to go, it was a place we had all been to a decade ago. Back then, it was serene, with a charm that made us feel connected to nature.

When we arrived, I realised how much had changed. The shore that used to be lined with simple huts and coconut trees now had tall buildings and fancy resorts. The roads were busy, and there were more shops and restaurants than I could remember. It was developed, yes, but it felt different, almost as if the place had lost a part of itself.

One evening during our vacation, we came across a notice about a turtle walk. It was an activity where volunteers would clean parts of the beach and help protect sea turtles. It sounded interesting, and we decided to join.

The walk began at dawn. The sky was in soft shades of orange and pink as the sun rose over the sea. The waves moved gently onto the shore, leaving patterns in the sand. The air smelled of salt, and there was a silence that made the moment feel special.

As we walked along the shore, picking up trash, I started noticing what we were collecting. There were plastic bottles, food wrappers, straws, and tiny bits of plastic that had broken down over time. They were everywhere, buried in the sand and floating in the shallow water. It felt wrong to see something so unnatural in a place so beautiful.

The sea looked endless, its colours shifting from deep blue to green. It seemed alive, constantly moving and breathing. The waves carried a rhythm that made me feel calm, almost as if the ocean itself was speaking. But beneath this beauty, there was a hidden problem. *Every piece of plastic we throw away eventually finds its way here, into the sea.*

I bent down to pick up a piece of a plastic bottle, covered in sand. It looked small and harmless, but I realised it wasn't. These bits of plastic don't go away; they stay, breaking into smaller and smaller pieces, harming the creatures that call the ocean their home.

As the turtle walk ended, I stood at the shore, looking out at the sea. It felt strange to think about how much we take from this beautiful place and how little we give back.

What is this plastic that has become such a big part of our lives, and why does it always end up here?

Plastic: A Double-Edged Sword

Plastic has become an integral part of our daily lives, offering convenience and versatility. Its invention revolutionized industries and solved numerous problems, but it has also led to significant environmental challenges.

The history of plastic dates back to the 19th century when Alexander Parkes introduced Parkesine in 1862, the first man-made plastic. This innovation was celebrated as it provided a substitute for natural materials like ivory and tortoiseshell, making products more accessible and reducing the exploitation of wildlife.

In the early 20th century, Bakelite, developed by Leo Baekeland in 1907, became the first fully synthetic plastic, marking the beginning of the modern plastics industry. Bakelite's versatility and durability led to its use in a wide range of products, from telephones to kitchenware, demonstrating plastic's potential to innovate across industries.

The production and use of plastics have grown exponentially since then. In 1950, the world produced around 2 million metric tons of plastic annually; by 2023, this number had surged to approximately 390 million metric tons.

Plastics have provided solutions in various sectors:

- **Healthcare**: Single-use plastic items, such as syringes and gloves, have improved hygiene and reduced the risk of infections.

- **Food Preservation**: Plastic packaging has extended the shelf life of perishable goods, reducing food waste.
- **Construction**: Lightweight and durable plastic materials have contributed to modern building techniques.

The very qualities that make plastic useful, its durability and resistance to degradation, have led to environmental problems. An estimated 20 million metric tons of plastic litter end up in the environment every year, affecting all land, freshwater, and marine ecosystems.

Plastic pollution has become ubiquitous in natural and built environments, raising concerns about potential harm to humans and nature alike. Once in the environment, research shows that plastic pollution is persistent and may take between 100 to 1,000 years or more to decompose, depending on environmental conditions.

The widespread use of plastics has also contributed to climate change. In the United States, plastic production is responsible for 232 million metric tons of greenhouse gases every year, equivalent to the emissions from 116.5 gigawatts of coal plants.

As I learned more about plastic, one thing became clear, there is something about single-use plastics that makes them hard to ignore. They are everywhere, from the bottles we drink out of to the bags we carry home. These plastics are designed to make life easier. You use them once, throw them away, and forget about them. But where do they really go after that?

Single-use plastics became popular in the mid-1900s when people started choosing convenience over reusable items. Plastic bags replaced cloth ones, disposable cutlery replaced metal, and plastic packaging became the norm for food. It felt practical and modern, and soon, these items became part of daily life.

Today, almost half of all the plastic produced in the world is for single-use items. Every year, millions of tonnes of these plastics are made, and much of it ends up as waste. A large portion finds its way into rivers, which carry it into the oceans. Studies show that there are already over 150 million tonnes of plastic floating in the sea, and single-use plastics are a big part of this.

The problem with single-use plastics is that they don't break down like natural materials. A plastic bottle or bag can remain in the environment for hundreds of years. Instead of disappearing, it breaks into tiny pieces called microplastics, which mix with the soil, water, and even the food we eat.

The impact on animals is heartbreaking. Sea turtles often mistake plastic bags for jellyfish and swallow them, which blocks their insides. Birds pick up small pieces of plastic, thinking they are food, and often die from eating too much waste. Fish, too, end up with tiny plastic bits in their stomachs, and these plastics make their way up the food chain to us.

On land, animals face similar challenges. They get tangled in plastic bags or eat wrappers, thinking they are edible. Over time, this waste spreads into every corner of the planet, our beaches, forests, rivers, and even the deepest parts of the oceans.

A Step Towards Hope

It made me think, if there are ways we are harming the planet, there must also be ways to protect it. While some of us are contributing to the problem, there are others working hard to find solutions. These are the people who remind us that progress does not have to mean destruction.

In Japan, researchers have created a new kind of plastic that breaks down quickly in water and soil. It does not take years to disappear or leave harmful chemicals behind. Instead, it dissolves in the sea within hours and enriches the soil with nutrients. It is made from safe, natural materials, meaning it is not just better for the environment but also useful for industries like packaging and medicine.

Another incredible innovation comes from scientists who are turning plastic waste into valuable products. They use a process that breaks down plastics like polystyrene into chemicals and clean energy. This method even produces hydrogen, which can be used as a renewable energy source. It is a way to recycle waste while also creating something new and helpful.

These solutions show that it is possible to fix the damage we have done. They are not perfect yet, but they are a step in the right direction. When I read about these efforts, it gives me hope. It reminds me that there are people out there who care deeply about the planet and are working to make a difference.

As I thought about the impact of plastics on the environment, it reminded me of how much we depend on other materials to build the world around us.

Building Our Cities, Impacting Our Planet

I have always been fascinated by the sight of tall buildings and bridges. They stand as symbols of progress, connecting cities and shaping the way we live. But the more I learn about the materials that make these structures, steel and cement, the more I understand the cost of creating them.

Steel and cement are everywhere. They make up the buildings we live in, the roads we drive on, and the bridges that span rivers. They are strong, durable, and essential to modern life. But they are also responsible for a significant part of the greenhouse gases that harm our planet.

Cement is a key ingredient in concrete, essential for construction. Its production involves heating limestone and other materials to high temperatures, a process that emits substantial amounts of carbon dioxide (CO_2). The cement industry alone is responsible for approximately 8% of global CO_2 emissions. This is due to both the energy-intensive nature of the production process and the chemical reactions involved.

Steel is no different. Steel is important for its strength and versatility in construction. Traditional steel production methods, such as blast furnaces, rely heavily on fossil fuels, leading to significant CO_2 emissions. On average, producing one ton of steel results in about two tons of CO_2 emissions. The steel industry contributes to roughly 7% of global CO_2 emissions, making it one of the largest industrial sources of greenhouse gases.

Together, these two materials are responsible for a large portion of the pollution caused by industrial production. They have helped us build cities and infrastructure, but they come with an environmental cost.

People are trying to change this. In steel production, methods like electric arc furnaces, which recycle old steel using electricity, produce fewer emissions. New ideas like green steel mills, which use hydrogen instead of coal, are being tested. A low-carbon steel mill planned in Brisbane, Australia, is expected to cut emissions by up to 80% for every tonne of steel it makes.

In the cement industry, alternatives are being explored too. Some factories are capturing the carbon dioxide released during production and storing it underground. Others are trying to use materials that produce less pollution when heated.

As I delved deeper into the environmental impact of construction materials like steel and cement, I encountered terms such as *Scope 1, Scope 2, and Scope 3 emissions.* These categories help us understand the different sources of greenhouse gases associated with industrial activities.

Scope 1 Emissions: Direct Emissions from Owned or Controlled Sources

Scope 1 emissions are the direct greenhouse gases released from sources that a company owns or controls. In the steel industry, for example, this includes emissions from burning fossil fuels in furnaces during steel production. Similarly, in cement manufacturing, the process of heating limestone to produce clinker emits CO_2 directly into the atmosphere. These emissions are significant contributors to a company's overall carbon footprint.

Scope 2 Emissions: Indirect Emissions from Purchased Electricity

Scope 2 emissions are the indirect emissions resulting from the generation of purchased electricity, steam, heating, or cooling consumed by the company. For instance, if a cement plant uses electricity from a coal-fired power plant, the associated emissions are considered Scope 2. Although the emissions occur at the power plant, they are attributed to the cement plant's energy consumption.

Scope 3 Emissions: All Other Indirect Emissions in the Value Chain

Scope 3 emissions encompass all other indirect emissions that occur in a company's value chain. This category is broad and includes emissions from various activities such as:

- **Purchased Goods and Services**: Emissions from the production of raw materials, like iron ore mining for steel manufacturing.
- **Transportation and Distribution**: Emissions from transporting raw materials to the factory and finished products to consumers.
- **Waste Generated in Operations**: Emissions from the disposal and treatment of waste produced during manufacturing.
- **Use of Sold Products**: Emissions resulting from the use of a company's products, such as the fuel consumption of vehicles made from steel.
- **End-of-Life Treatment of Sold Products**: Emissions from the disposal or recycling of products at the end of their life cycle.

Scope 3 emissions often represent the largest portion of a company's total greenhouse gas emissions, making them important to address for comprehensive carbon management.

By identifying the sources of emissions, businesses can implement targeted strategies to mitigate them, such as improving energy efficiency, transitioning to renewable energy sources, and engaging with suppliers to reduce upstream emissions.

As I learned more about Scope 1, 2, and 3 emissions, it reminded me of conversations I've had during my work with industries. Reducing emissions is not easy, but it is something many companies are trying to do. Every industry has its own challenges, and the solutions are rarely straightforward.

I've seen businesses take small but meaningful steps. Some factories are switching to renewable energy, like solar or wind, to power their operations. Others are improving their machines and equipment so they use less fuel and work more efficiently. These changes help reduce direct emissions, the kind that come straight from running factories.

Then there are companies working with their suppliers to make a bigger impact. They ask suppliers to use cleaner methods for mining raw materials or producing components. It's a longer process, but it helps reduce emissions from the supply chain.

Some industries are even looking at how their products are used after they leave the factory. For example, car manufacturers are developing electric vehicles to reduce emissions during use. Others are designing products that can be easily recycled, so less waste ends up in landfills.

The biggest challenge is measuring emissions accurately. Many companies struggle to track what happens in their supply chains or how their products are used and disposed of. But there is progress. New tools and technologies are making it easier to gather data and find solutions.

The most interesting thing I've noticed is that the change starts small. A company might begin with one cleaner machine, one solar panel, or one discussion with a supplier. But those small steps grow into larger ones.

It makes me wonder, if every industry took those first steps, how far could we go?

Chapter 5:

Global Climate Agreements – Promises and Progress

Travelling was a part of my job. This time, I needed to collect some reports in Delhi. I packed my bags, adding a few extra sweaters since it was winter. I remember visiting Delhi years ago, a beautiful city with markets and open spaces.

As the plane approached Delhi, I looked out, expecting to see the familiar sights. Instead, everything seemed blurred. The pilot announced that we couldn't land immediately due to low visibility caused by fog and pollution. We circled for what felt like hours.

When we finally landed, the air felt different. People wore masks, and there was a faint smell in the air. I learned that Delhi's Air Quality Index (AQI) had reached alarming levels, with readings soaring past 450 on some days, even touching 800 in certain areas. Anything above 300 is considered hazardous.

The drive into the city was unsettling. Schools were closed, construction activities halted, and residents were advised to stay indoors.

The streets, which were always full of people, now felt silent. Famous landmarks were hard to see in the fog.

In January 2025, Delhi recorded its warmest January day in six years, with temperatures reaching 26.1°C. Despite the warmth, the air quality remained in the 'very poor' category, with an AQI of 349.

The Indian Institute of Tropical Meteorology's Air Quality Early Warning System reported that on January 18, 2025, at 4 PM, Delhi's AQI was 255, placing it in the 'poor' category.

A report by the Energy Policy Institute at the University of Chicago highlighted that Delhi residents are losing 7.8 years of life expectancy due to elevated PM2.5 levels.

Despite various measures, including the Graded Response Action Plan (GRAP), air quality remains a significant concern. The city's rapid urbanization, industrial activities, and vehicular emissions continue to challenge efforts to improve air quality.

As I moved through the city, memories of the Delhi I once knew contrasted with the reality before me. It wasn't the same place I had visited 15 years ago. *I couldn't help but wonder, what is this pollution, and how did it change the city so drastically?*

As I thought about Delhi's air pollution, it reminded me of the larger picture, how countries are working together to address climate change. One of the most significant steps in this direction is the Paris Agreement. It's a plan that brings nations together with a shared goal, to protect the planet from the worst effects of global warming.

The Paris Agreement: A Global Effort to Combat Climate Change

The Paris Agreement was adopted in December 2015 at the United Nations Climate Change Conference (COP21) in Paris. Its main goal is to keep global warming well below 2°C compared to pre-industrial levels, with efforts to limit it to 1.5°C. Scientists have warned that crossing the 2°C threshold would lead to severe and irreversible damage, including rising sea levels, extreme weather, and loss of biodiversity.

Under the agreement, countries committed to reducing greenhouse gas emissions to slow down global warming. Each nation developed its own climate action plan, known as Nationally Determined Contributions (NDCs). These plans outline how they will cut emissions and shift to cleaner energy sources. The idea was to let countries set their own targets based on their circumstances, while still working towards the global goal.

The Paris Agreement encouraged countries to move away from fossil fuels like coal and oil, which are major sources of greenhouse gases. Instead, many nations have committed to increasing their use of renewable energy, such as solar, wind, and hydroelectric power.

India has taken major steps in this direction. By 2030, the country plans to generate 50% of its electricity from renewable sources. Solar power plants and wind farms are being built across the country. As of 2023, India's renewable energy capacity has grown to over 120 gigawatts, placing it among the top producers of clean energy in the world. Rooftop solar systems are also being promoted, allowing households and small businesses to contribute to the renewable energy transition.

The European Union has set a goal to reduce emissions by 55% by 2030 compared to 1990 levels. To achieve this, many EU nations are increasing investments in wind and solar energy. Germany, for instance, has established large solar parks and onshore wind farms, while Denmark has focused on offshore wind energy. These efforts are supported by research into advanced renewable technologies, which aim to make energy production more efficient.

In the United States, the government has committed to achieving 100% clean electricity by 2035. Meanwhile, China, which is the largest emitter of greenhouse gases, is rapidly expanding its renewable energy projects. By 2025, China plans to add 1,200 gigawatts of solar and wind power, a massive step toward reducing its reliance on coal.

This transition to renewable energy is helping countries reduce emissions while supporting economic growth. Solar and wind industries are creating jobs in installation, manufacturing, and maintenance. The costs of renewable energy have also dropped significantly in the past decade, making clean energy more accessible and affordable.

One part of the Paris Agreement that stood out to me was its focus on helping countries adapt to the impacts of climate change. Reducing emissions is important, but some countries are already facing problems like floods, droughts, and rising sea levels. These challenges are making life harder, especially for nations that lack resources to protect themselves.

To support these vulnerable countries, the Paris Agreement encourages financial and technical aid. Developed countries have pledged $100 billion annually to help fund projects that address climate-related issues. This money goes towards building stronger defenses against floods, improving water systems in areas affected by drought, and restoring lands damaged by environmental changes.

For small island nations, rising sea levels are a constant threat. Many are using this support to create coastal barriers and protect their communities from the ocean. In places where droughts are common, the funds are used to set up irrigation systems and introduce crops that need less water to survive.

Another focus is on preparing for extreme weather. Early warning systems are being introduced in areas prone to disasters like cyclones and floods. These systems give people time to prepare, evacuate if necessary, and reduce the loss of lives and property. Communities are also being trained to respond quickly during such events.

The agreement also pushes for collaboration between countries. Sharing knowledge and technology is a key part of this effort. Some nations are helping others adopt renewable energy systems, while others are sharing farming techniques to ensure food security in difficult climates.

As I read about these measures, it became clear how deeply connected we all are. The challenges may be different for each region, but the solutions require everyone to work together. It's a global effort to protect what we have and prepare for the changes ahead.

Reflecting on global efforts to combat climate change, I recalled one more conference.

COP26

The 26th United Nations Climate Change Conference of the Parties, known as COP26, took place in Glasgow, Scotland, from 31 October to 13 November 2021. This annual conference brings together countries to discuss and negotiate actions to address climate change.

At COP26, nearly 200 nations participated, aiming to strengthen commitments made under the Paris Agreement of 2015. The primary focus was on limiting global warming to well below 2°C, with efforts to keep it to 1.5°C above pre-industrial levels.

One of the significant outcomes was the Glasgow Climate Pact, which emphasized the need for countries to present more ambitious climate action plans. This included accelerating efforts to phase down unabated coal power and inefficient fossil fuel subsidies.

Additionally, developed countries were urged to scale up climate finance, aiming to collectively mobilize $100 billion per year to assist developing nations in mitigating and adapting to climate impacts.

COP26 also addressed the importance of protecting and restoring ecosystems, recognizing the role of nature in achieving climate goals. This included commitments to halt and reverse deforestation and land degradation by 2030.

The conference highlighted the urgency of immediate and collective action, with countries acknowledging that current efforts were insufficient to meet the targets set by the Paris Agreement. The Glasgow Climate Pact called for nations to revisit and strengthen their 2030 emission reduction targets by the end of 2022.

Reading about the Paris Agreement and COP26 gives me hope. It feels like the world is starting to take real steps to address climate change. But seeing the reality in a city like Delhi makes it clear how much work is still left to do.

I had finished my work in the city and was flying back home. As I sat on the plane, the memories of the trip stayed with me. Delhi felt so different from the last time I had visited. The roads were packed with vehicles, and the air felt far from clean. Masks were everywhere, warnings about the air quality flashed on signs, and it all seemed like a constant reminder of how much had changed.

The plane began to take off, and I looked out of the window. Everything below seemed faded, as though the city was hidden under a layer of smoke. I thought about how far we've come as a world, building cities and growing economies. Yet, here we are, struggling to breathe in those very cities.

The air we breathe is something we all share, and it doesn't stop at borders. Sitting on that flight, I couldn't stop asking myself, how did we let it come to this?

Chapter 6:

Heavy Industries – Challenges and Opportunities

The plan was simple, trek during the day, camp at night, and wake up to fresh air. It sounded perfect. Some exercise, a break from the usual routine, and a chance to be in nature. I packed my bag, added an extra sweater for the night, and joined the group.

The trek started early. The path was surrounded by trees, and the air felt fresh. Everyone walked at their own pace, stopping now and then to take a sip of water or stretch their legs. There was no rush, and that made it better. Walking through the forest, hearing the sounds around, it felt like a good way to spend the day.

By evening, we reached the campsite. The tents were set up, and soon, a fire was lit. Sitting around it, eating simple food, and talking to people I had just met felt easy. The warmth of the fire felt good in the cold. The night passed with conversations, stories, and laughter.

Later, we got into our tents. The ground was uneven, and the sleeping bag wasn't as soft as I thought it would be, but it didn't matter. The day had been long, and sleep came without much effort.

Morning felt different. The air was fresh, and everything around looked untouched. It felt good to wake up to that. Packing up didn't take long, and soon, we were heading back. The trek down felt lighter, and the conversations were easier.

Back home, I stepped in through the gate and picked up the newspaper. The first thing I saw, **Los Angeles, Wildfire Disaster.**

The news detailed massive wildfires that had spread across Los Angeles County. The Palisades and Eaton fires had destroyed entire neighborhoods, leaving thousands of families without homes. The fires had started on **January 7, 2025**, tearing through the dry land and spreading faster than firefighters could control.

The Palisades Fire burned through parts of **Pacific Palisades and Malibu**, destroying over **6,800 buildings** and burning more than **23,000 acres** of land. The Eaton Fire, which spread through **Altadena and Pasadena**, was even worse, burning over **14,000 acres** and damaging at least **9,400 buildings**. The numbers alone felt overwhelming, but the reality of the damage was worse.

The cause of the fires was still under investigation, but reports pointed to multiple factors. **Southern California Edison** revealed that its power lines had recorded an unusual spike in current in the Eaton Canyon area just before the fire started. Investigators were looking into whether a faulty transmission line could have caused sparks that led to the disaster.

The conditions in California had made the situation worse. A report showed that rising temperatures, low rainfall, and dry air had extended the fire season, making these disasters more common. Areas that once faced high fire risks for a few months were now at risk for most of the year.

The cost of the fires was enormous. Insurance companies estimated the total losses to be over **$28 billion**, making it one of the most expensive disasters in U.S. history. Businesses were struggling, with small restaurants and stores forced to shut down. A popular seafood restaurant in Malibu reported losing nearly **$750,000** in damages and lost sales, struggling to reopen.

I sat down, reading through the pages. A night by a campfire felt like a luxury when, somewhere else, an uncontrollable fire was destroying lives. The contrast felt strange, one moment, fire was something to gather around, and in another, it was tearing through homes and turning cities to ash.

The more I read, the more I thought about where this was leading. The increasing number of wildfires, the extreme heat, the damage to communities, none of this was happening on its own. The industries that power our cities, produce the materials we use, and keep economies running all play a role.

Where do we go from here? That's the real question.

Heavy Industries

Fires don't happen on their own. The world has changed in ways that make them worse, hotter summers, drier land, and stronger winds. The way we build, produce, and consume has shaped this reality.

Everywhere we look, we are surrounded by materials made by heavy industries. The buildings we live and work in, the roads we drive on, and the bridges that connect cities all come from industries like steel, cement, and chemicals. These industries keep everything running. Without them, there would be no skyscrapers, highways, or factories.

Steel is in cars, railways, and construction. Cement holds together the structures we rely on every day. Chemicals are used in fertilisers, medicines, plastics, and cleaning products. These industries provide jobs and support economies worldwide.

At the same time, they are some of the biggest sources of carbon dioxide and other greenhouse gases. The steel industry alone contributes around **7% of global carbon dioxide emissions**, while cement production is responsible for about **8%**. The chemical industry, which produces plastics, fertilisers, and other essential materials, adds even more to the total.

These industries rely on processes that burn large amounts of coal, oil, and gas. Making steel requires heating iron ore at extremely high temperatures, usually using coal-fired blast furnaces. Cement production releases emissions both from burning fuel and from the chemical reaction itself. The chemical industry depends on fossil fuels to produce many of its raw materials, making it one of the hardest sectors to decarbonise.

For years, these emissions have been treated as an unavoidable part of industrial progress. The demand for steel, cement, and chemicals keeps growing, and with that, so does the environmental impact.

How do we keep building and producing without making the problem worse?

Innovative Solutions

1. Capturing Carbon Before It Reaches the Atmosphere

Thinking about how much steel, cement, and chemicals contribute to greenhouse gas emissions made me wonder, what can actually be done to reduce this? These industries are responsible for a large share of pollution, and cutting emissions from them is not as simple as switching to renewable energy. One solution that has been gaining attention is **Carbon Capture and Storage (CCS)**.

CCS is a process that captures carbon dioxide before it escapes into the air and stores it safely underground. This method is being developed to reduce emissions from factories, power plants, and industrial sites where cutting pollution is difficult. Since industries like steel, cement, and chemicals burn huge amounts of fuel, CCS helps by stopping these emissions at the source.

The process happens in three steps, **capturing the carbon**, **transporting it**, and **storing it safely**.

Capturing Carbon Dioxide

Factories and power plants release carbon dioxide when they burn fuels like coal, oil, or gas. CCS captures these emissions before they escape. There are three main ways to do this:

- **Post-Combustion Capture** – This method removes carbon dioxide from the gases released after burning fuel. Factories use chemical solutions that absorb the carbon, separating it from the other gases. This method is already being used in some coal and gas power plants.
- **Pre-Combustion Capture** – Instead of waiting until after burning fuel, this method captures carbon before the combustion process. Fuels like coal or gas are turned into a mixture of hydrogen and carbon dioxide. The carbon is then separated, leaving hydrogen behind, which can be used as a cleaner fuel.
- **Oxy-Fuel Combustion** – Here, fuel is burned in pure oxygen instead of air. This creates waste gases made up of mostly carbon dioxide and water vapour. The water is removed, leaving nearly pure carbon dioxide ready for storage.

These methods can capture up to **90% of the carbon dioxide** produced at an industrial facility, significantly reducing emissions.

Transporting Carbon Dioxide

Once the carbon dioxide is captured, it needs to be moved to a storage site. The gas is compressed into a liquid-like state to make transportation easier. It is then sent through **pipelines**, **ships**, or **trucks** to locations where it can be safely stored underground.

Pipelines are the most common way to transport carbon dioxide over long distances. Many countries are already building **carbon transport networks**, connecting industries to storage sites. Ships are used when pipelines are not available, especially in places where carbon needs to be transported overseas.

Storing Carbon Underground

The final step is **long-term storage**. Instead of releasing carbon dioxide into the air, it is pumped deep underground into **rock formations that can hold it safely for thousands of years**. These storage sites include:

- **Depleted Oil and Gas Fields** – These are underground reservoirs that once held oil and gas. Since they kept those fuels trapped for millions of years, scientists believe they can safely store carbon dioxide as well.
- **Deep Saline Aquifers** – These are underground rock layers filled with saltwater. The carbon dioxide is injected into these layers, where it dissolves in the water and gets locked away.

- **Unmineable Coal Seams** – Some deep coal layers cannot be mined, but they can store carbon dioxide. In some cases, injecting carbon into these seams also helps release methane gas, which can be used as an energy source.

Scientists continuously monitor these storage sites to ensure that the carbon dioxide stays trapped and does not leak back into the air.

Where Is It Being Used?

When I first read about Carbon Capture and Storage (CCS), it sounded like a futuristic idea, trapping carbon dioxide before it enters the air and storing it deep underground. But the more I looked into it, the more I realised that this isn't some experiment for the future. It is already happening in several countries, and the scale at which it is being used is much larger than I expected.

One of the earliest and most successful CCS projects is in **Norway**. The **Sleipner Project**, which has been operating since **1996**, is the world's first large-scale CCS facility. It captures **about 1 million tonnes of carbon dioxide every year** from natural gas production and stores it deep under the **North Sea**. The carbon is injected into a sandstone reservoir, preventing it from being released into the air. This project alone has helped Norway set an example for how CCS can work in real conditions.

Norway didn't stop there. Another major CCS facility, **Snøhvit**, has been running since

2008. It captures and stores **700,000 tonnes of CO_2 per year** from a liquefied natural gas (LNG) plant. Unlike many other CCS projects, Snøhvit is unique because it is **offshore**, meaning that all carbon capture and storage happens beneath the seabed.

In the **United States**, one of the biggest CCS projects was **Petra Nova in Texas**. It started in **2017** and captured **over 4 million tonnes of CO_2** before shutting down in **2020** due to financial issues. The facility used captured carbon to extract more oil from existing wells, a process known as **enhanced oil recovery (EOR)**. Though it proved that CCS works on a large scale, the economic challenges showed that making it cost-effective is still a problem.

Another successful CCS project in **North America** is **Boundary Dam** in **Canada**. This coal-fired power plant in **Saskatchewan** was the first in the world to integrate CCS on a commercial scale. Since it began operating in **2014**, it has captured over **1 million tonnes of CO_2 per year** and has been a model for future projects. Unlike Petra Nova, Boundary Dam continues to operate and has helped shape CCS policies in Canada.

There are also newer projects being developed across different industries. **The Gorgon Project** in **Australia**, which started in **2019**, aims to capture **4 million tonnes of CO_2 per year** from a natural gas facility. Meanwhile, **China**, the world's largest emitter of greenhouse gases, has been investing in CCS as well, with multiple pilot projects in operation.

Why Is It Important?

Reading about these large-scale CCS projects made me think about something bigger. This is about keeping industries running while reducing the damage they cause. Heavy industries like steel, cement, and chemicals cannot stop overnight. The demand for these materials is high, and the world depends on them. Carbon Capture and Storage is one way to lower emissions while industries continue to operate.

For industries that rely on fossil fuels, switching to clean energy takes time. Power plants can use solar or wind, but factories that produce cement or steel need extremely high temperatures that are difficult to achieve without coal or gas. CCS helps by **trapping emissions before they enter the atmosphere**, allowing these industries to continue production while reducing pollution.

Many experts believe that CCS is necessary to lower global emissions. Reports suggest that without it, keeping global warming below **1.5°C** will be extremely difficult. The **Intergovernmental Panel on Climate Change (IPCC)** has included CCS as a solution for industries that are harder to decarbonise. Some studies estimate that by **2050**, CCS could prevent up to **15%** of total global emissions.

Large projects exist, but most industries have not adopted CCS at full scale. Cost remains a major challenge, with some plants shutting down because they are too expensive to operate. Governments and private companies are investing in research to make it more efficient and affordable, but progress is slow.

Carbon Capture and Storage is one possible way forward. It has worked in multiple locations, but is it enough to solve the problem? That is the real question.

2. Green Hydrogen: Powering Factories with Clean Energy

After learning about Carbon Capture and Storage, I became curious about other ways to reduce emissions in heavy industries. This led me to discover **green hydrogen**, a clean fuel that could transform how factories operate.

What Is Green Hydrogen?

Green hydrogen is made by splitting water into hydrogen and oxygen using electricity from renewable energy sources like wind and solar power. This process is called **electrolysis**. Since no fossil fuels are used, green hydrogen does not release carbon dioxide, making it a clean energy source.

Hydrogen is already used in industries, but most of it comes from natural gas, which produces emissions. Green hydrogen offers a way to use hydrogen without adding pollution to the air. This makes it an important option for industries that need a reliable energy source but want to reduce their emissions.

Industries like steel, cement, and chemicals need extremely high temperatures for their production processes. They usually burn coal, oil, or natural gas to generate the heat required. Green hydrogen can replace these fuels, allowing factories to operate while reducing emissions.

Steel Production

Steel is usually made using coal in blast furnaces. This process releases large amounts of carbon dioxide. Green hydrogen can replace coal in a method called **direct reduction**, where hydrogen is used to remove oxygen from iron ore. This produces water instead of carbon dioxide.

The **H2 Green Steel Project** in Sweden is working to produce steel using hydrogen instead of coal. If successful, this method could remove a huge amount of emissions from the steel industry, which contributes about **7% of global carbon dioxide emissions**.

Cement Manufacturing

Cement production is responsible for about **8% of global carbon dioxide emissions**. Some of these emissions come from burning fossil fuels to heat limestone. Green hydrogen can be used instead to provide the necessary heat. Companies like **Hanson UK** are testing hydrogen-powered kilns to reduce emissions from cement factories.

Chemical Industry

The chemical industry depends on hydrogen for making fertilisers, plastics, and other products. Most of this hydrogen comes from natural gas, which produces carbon dioxide. Green hydrogen can be used instead, removing the emissions linked to chemical production.

One of the biggest uses of hydrogen is in **ammonia production**, which is a key ingredient in fertilisers. The **Haber-Bosch process**, which combines hydrogen and nitrogen to make ammonia, has been used for over a century. Now, companies are working to replace fossil-fuel-based hydrogen with green hydrogen to cut down emissions.

Current Developments and Future Prospects

Countries around the world are investing in green hydrogen to reduce industrial emissions. Governments and private companies are working on projects that aim to replace fossil fuels with this clean energy source. Some of the biggest developments are happening in **Europe, Australia, and China**, where large-scale hydrogen production and transport networks are being set up.

Europe's Green Hydrogen Projects

Europe has taken the lead in developing green hydrogen as part of its plan to cut emissions from industries and transport. Several projects have been launched to expand hydrogen production and build the infrastructure needed to distribute it.

- **The Green Spider and Green Flamingo Projects:** These initiatives aim to connect hydrogen production sites with industries that need clean energy. The projects focus on building pipelines and storage facilities to move green hydrogen efficiently across different regions.

- **Germany's Hydrogen Strategy**: Germany is investing **€9 billion** to scale up hydrogen production. The country is working on partnerships with other nations, including Norway and Morocco, to import green hydrogen for its industries.
- **The Netherlands' Offshore Hydrogen Hub**: The country is planning to use its offshore wind farms to produce hydrogen at sea. This will help power industries and reduce dependence on coal and natural gas.

Australia's Large-Scale Hydrogen Plans

Australia is building some of the world's biggest green hydrogen plants, using its strong renewable energy resources to power the production process.

- **The Western Green Hydrogen Hub**: Located in Western Australia, this project aims to produce **up to 3.6 million tonnes of green hydrogen per year**. The hydrogen will be used for industries and exported to countries looking for clean energy sources.
- **The Asian Renewable Energy Hub**: Another major project in Western Australia that combines large-scale wind and solar farms to produce green hydrogen for export.

- **Queensland Hydrogen Taskforce**: Australia's east coast is also seeing investment in hydrogen production, with plans to supply local industries and export hydrogen to Japan and South Korea.

China's Push for Green Hydrogen

China, the world's largest producer and consumer of hydrogen, is shifting its focus towards green hydrogen to cut emissions from heavy industries and transport.

- **Hydrogen Cities Pilot Program**: China is developing multiple hydrogen hubs in cities like Beijing and Shanghai, where industries and vehicles will be powered by green hydrogen.
- **State-Owned Hydrogen Projects**: Chinese companies, including **Sinopec**, are investing in large-scale green hydrogen production. Sinopec's new project in Inner Mongolia is expected to produce **30,000 tonnes of green hydrogen annually**.
- **Integration with Renewable Energy**: China is linking its solar and wind farms to hydrogen production facilities to ensure that the hydrogen is generated without fossil fuels.

Green hydrogen has the potential to replace fossil fuels in industries, but large-scale production is still expensive. Producing hydrogen through electrolysis requires a lot of electricity, and renewable energy capacity needs to grow to support widespread hydrogen use. Governments and companies are investing in new technology to make hydrogen production more affordable and efficient.

The world is moving towards using green hydrogen in industries and transport. Projects are growing in size, and investment is increasing. The question now is whether green hydrogen can become available at a scale that truly makes a difference.

3. Circular Economy: The Importance of Recycling and Efficient Resource Use

After exploring green hydrogen as a clean energy solution, I began to think about other ways we can make our industries and lifestyles more sustainable. This led me to the concept of the **circular economy**, which focuses on recycling and using resources efficiently to reduce waste and environmental impact.

In a traditional linear economy, we follow a *"take-make-dispose"* model: we extract resources, create products, use them, and then throw them away. This approach leads to a lot of waste and depletes natural resources. In contrast, a circular economy aims to keep products and materials in use for as long as possible. This means designing products that can be repaired, reused, or recycled, and finding ways to turn waste into new resources.

The Role of Recycling

Recycling is a key part of the circular economy. By collecting and processing materials that would otherwise be discarded, we can reduce the need for new raw materials and decrease pollution. For example, recycling aluminium saves up to **95% of the energy** required to produce the same amount from raw materials. Similarly, recycling paper reduces energy consumption and conserves trees.

Recycling rates vary widely depending on the material and region. According to a report by the **International Resource Panel**, less than a third of the 60 metals studied have an end-of-life recycling rate above 50%, and 34 of them have a recycling rate below 1%. This shows that there is significant room for improvement in our recycling efforts.

Beyond recycling, the circular economy emphasizes using resources more efficiently throughout their lifecycle, from extraction and manufacturing to consumption and disposal. This includes designing products that last longer, can be easily repaired, or are made from materials that can be fully recycled. It also involves rethinking business models to focus on services rather than products, such as offering product-as-a-service models where consumers rent or lease products instead of owning them.

Many countries and organisations are recognizing the benefits of a circular economy and are taking steps to implement its principles. For instance, the **European Union** has developed a comprehensive monitoring framework to measure progress towards a circular economy, focusing on areas like production, consumption, waste management, and secondary raw materials.

In **Australia**, the government has launched a Circular Economy Framework aiming to significantly enhance material reuse, repair, and recycling to decrease waste and environmental harm while boosting the economy. The initiative aims to add $26 billion to the economy and divert 26 million tonnes of waste from landfills each year by doubling the country's "circularity" rate from the current 4.6% to closer to the global average of 7%.

Transitioning to a circular economy is not without challenges. It requires changes in consumer behaviour, redesigning products and processes, and developing new business models. Additionally, robust data and indicators are needed to measure progress and inform policy decisions. The **United Nations Environment Programme** emphasizes the importance of developing such indicators to assess the effectiveness of circular economy initiatives.

Recycling Waste into Something Useful

Cities are growing fast, and with them, the amount of waste keeps increasing. Old vehicles sit in scrapyards, piling up. Garbage from homes and businesses fills landfills, releasing harmful gases. But there are ways to turn this waste into something useful.

Many cars, trucks, and buses that are no longer safe to drive end up as scrap. They take up space, cause pollution, and add to the problem of waste. Instead of letting them rust away, their steel can be recycled and used again.

India's Vehicle Scrappage Policy is encouraging people to retire old, polluting vehicles. This does two things. First, it removes cars that release high levels of emissions. Second, it provides scrap metal, which can be melted down and reused. This recycled steel, known as green steel, requires less energy to produce than steel made from raw materials. It also reduces the need for mining iron ore, which damages the environment.

Steel factories are now using more scrap to lower emissions. Producing green steel from recycled metal can cut pollution by up to 60%. Countries like Germany and Sweden are already leading the way, and India is beginning to follow. The more scrap steel is reused, the less we rely on traditional steel production, which burns large amounts of coal.

Turning Municipal Waste into Fuel

Another growing problem in cities is the amount of rubbish thrown away every day. Most of it ends up in landfills, where it takes years to decompose. But some of this waste can be turned into fuel.

Refuse-Derived Fuel (RDF) is made from non-recyclable waste like plastics, paper, and leftover materials. Instead of dumping these into landfills, they are processed and used as fuel for cement factories and power plants. This reduces the need for coal and lowers emissions.

Many cement plants in India are already using RDF to fire their kilns. This not only helps manage waste but also provides an alternative energy source. It is a way to reduce landfill waste while keeping industries running. Some countries have taken this even further, using waste-to-energy plants to generate electricity from materials that cannot be recycled.

Recycling old vehicles into green steel and turning waste into fuel are practical ways to reduce pollution and save resources. These are not futuristic ideas, they are already happening. If these efforts expand, they could help solve two problems at once: reducing waste and lowering emissions. The challenge is making these solutions more common so they become a regular part of how industries work.

Thinking about the circular economy made me realise something, nothing in nature goes to waste. Everything has a purpose, and what is discarded in one place becomes useful somewhere else. That is how the planet has always worked. The challenge now is whether we can do the same with the things we produce, use, and throw away.

Recycling, reusing, and making better use of resources can reduce the pressure on industries that burn through raw materials and release emissions. If less waste is created, there is less need to extract, process, and transport new materials. It is a simple idea, but putting it into action on a large scale is where the real challenge lies.

As I sat with these thoughts, I was reminded of the campfire from the night in the forest. The way we sat around it, warming our hands, sharing stories, and letting the flames bring comfort in the cold. Fire is something that can bring people together. It can provide warmth and light, but when it is out of control, it destroys everything in its path.

I turned to the newspaper again and looked at the photographs of the wildfires in Los Angeles. Entire communities had been lost. Homes turned to ashes. The same fire that brings warmth and light can take away lives when the balance is lost.

It made me wonder, how much of what we see as progress is actually pushing us closer to destruction?

Chapter 7:

The Sustainable Development Goals (SDGs) – A Global Blueprint

Birds must have it easy. No honking, no waiting at signals, no getting stuck behind a long queue of cars. Just open space, flying wherever they want. I thought about that while planning a trip to Uttarakhand. It had been a while since I had gone somewhere new, and a visit to a bird sanctuary sounded like a good way to spend the weekend.

Travelling always has a way of making things interesting. The roads winding through hills, small towns with tea stalls at every turn, and the sight of people going about their day. The journey itself was something to enjoy. By the time I reached the sanctuary, I was ready to stretch my legs and see the birds.

The place was supposed to be full of migratory birds this time of year. They come from Europe and Central Asia, escaping the cold and staying till March before heading back. That is how it had always been. But something felt different.

I walked around, expecting to see flocks resting by the water or flying in groups. There were a few birds here and there, but nowhere near as many as I had imagined. I asked the guide about it, and he shook his head. *"They left early this year,"* he said.

That did not make sense. These birds have been coming here for generations, always following the same pattern. But this time, they had flown back weeks earlier than usual. February should have still been winter, but the temperatures had **crossed 20°C**, about **4°C higher** than last year. That was not the kind of weather they were used to. The heat had changed their schedule.

I kept thinking about it as I made my way back. Birds do not check weather reports or make backup plans. They just know. They sense the smallest shifts in temperature, food availability, and wind patterns. If something is off, they move. That is how nature works.

I looked out of the car window as the city came into view. Buildings, roads, people, cars, all moving in different directions. Everything built to make life easier, to make things faster. But when things change too fast, do we know when to move?

That is where **The Sustainable Development Goals (SDGs)** come in. They were made to prepare for a future that is already arriving. *But how far have we come?*

I realised that such changes in nature are signals of broader global challenges. To address these issues, the United Nations established the **Sustainable Development Goals (SDGs)** in 2015. These **17 goals** provide a comprehensive plan.

These goals cover many things, poverty, clean water, education, energy, and protecting nature.

One of the important goals is **climate action**. Rising temperatures, unpredictable weather, melting glaciers, and extreme weather events are already affecting lives. The SDGs focus on reducing emissions, using more clean energy, and protecting forests and oceans. But climate change is not something one country can solve alone. It affects everyone, and it needs countries to work together.

Another key goal is **sustainable cities and communities**. Cities are growing fast, and if they are not planned well, they become major sources of pollution. This goal focuses on making cities better by reducing waste, using clean energy, and improving public transport.

These goals were made to prepare for the future. But with everything happening around us, rising temperatures, floods, wildfires, and changing seasons, the real question is, **are we acting fast enough?**

SDG 7: The Need for Affordable and Renewable Energy

Energy is a part of everyday life. It powers homes, industries, and transportation. The need for energy keeps increasing, but at the same time, the way it is produced has a serious impact on the planet. Recognising this, the **United Nations introduced Sustainable Development Goal 7 (SDG 7), which**

focuses on ensuring access to affordable, reliable, sustainable, and modern energy for everyone by 2030.

Over the years, there has been steady progress in providing electricity across the world. In 2000, around **78% of the global population had electricity**. By 2021, this increased to **91%**. This progress has been driven by significant developments in Asia, especially in countries like India and China. South Asia alone went from **58% access to 99% in two decades**. However, **Sub-Saharan Africa remains the biggest challenge,** where access has improved from **26% to 51%,** but millions still rely on outdated and polluting energy sources. If things continue at the current pace, **around 660 million people will still be without electricity by 2030**.

Electricity access is one side of the challenge. The other is clean cooking solutions. Around **2.3 billion people**, nearly **one-third of the world's population**, still use traditional fuels like wood and charcoal for cooking. This leads to dangerous indoor air pollution, which is linked to millions of deaths every year. If the pace of change does not pick up, the world will fall short of its clean cooking target by about **30%** by 2030.

Transitioning to Renewable Energy

One of the biggest changes happening in energy is the shift towards renewables. Instead of depending on coal, oil, and gas, more countries are turning to solar, wind, and hydropower.

- **Solar Power**: Over the past decade, the cost of solar panels has dropped sharply. A decade ago, solar energy was expensive, but today, the cost has come down by around **85%**, making it one of the most affordable energy sources. The amount of solar power installed worldwide has increased from **42 GW in 2010 to 714 GW in 2020**.
- **Wind Energy**: Wind power is now one of the cheapest ways to generate electricity. Countries with good wind resources, like the UK, Denmark, and Germany, have built offshore and onshore wind farms to power homes and industries.

In 2016, **renewables made up 17.5% of the world's total energy consumption**. This share has been rising as more investments go into solar, wind, and hydroelectric power.

Using energy efficiently is just as important as producing clean energy. The aim is to **double the rate of improvement in energy efficiency by 2030**. This means reducing the amount of energy used for every unit of economic output.

Some countries have taken major steps to improve efficiency by switching to LED lighting, using smart grids, and promoting electric transport. Governments are also offering financial incentives to encourage industries and households to use energy-efficient appliances.

Shifting to clean energy requires massive investments. In developing countries, international funding for renewable energy has increased significantly since 2010. Most of the funding has gone into **hydropower (50%) and solar projects (20%)**. Expanding infrastructure and upgrading technology are necessary to provide energy access in remote regions, especially in small island nations and landlocked countries.

The Climate Connection

Energy production is one of the largest contributors to greenhouse gas emissions, making up about **60% of global emissions**. The burning of fossil fuels such as coal, oil, and gas releases large amounts of carbon dioxide into the air. This has led to rising temperatures, extreme weather events, and environmental damage. It has also affected human health, with air pollution linked to serious illnesses, including lung diseases and heart conditions.

Renewable energy sources such as solar and wind provide a way to generate electricity without releasing harmful emissions. These sources are already being used in many countries to reduce dependence on fossil fuels. A study from the **Harvard T.H. Chan School of Public Health** found that switching to renewable energy and using energy-saving technologies could prevent many health problems caused by air pollution. Some estimates show that moving away from fossil fuels could save millions of lives by reducing air pollution-related deaths.

There has been progress in providing electricity to more people and expanding the use of renewables, but the changes are happening too slowly. If the world is to meet its goals by 2030, much more needs to be done. Every delay makes the effects of climate change harder to manage.

As temperatures rise and extreme weather worsens, cities play a huge role in the fight against climate change. Home to over half the world's population, they consume massive energy and produce significant carbon emissions. Are we acting fast enough to turn things around?

To make sure cities remain liveable, **Sustainable Development Goal 11 (SDG 11)** was introduced. It focuses on making cities safer, more sustainable, and better prepared for climate-related challenges.

By **2050**, nearly **70% of people** in the world will be living in cities. Right now, this number is **55%**, and it is growing fast. This puts pressure on housing, transport, and public services. The demand for space means cities are expanding, often without proper planning.

More than **1.1 billion people** live in slums or poor housing conditions. These areas are at high risk of climate-related disasters. When floods or heatwaves occur, they are often the most affected. Cities occupy only **3% of the Earth's land**, yet they use **60–80% of the world's energy** and produce **75% of carbon emissions**. Urban expansion is leading to more air pollution, loss of green spaces, and inefficient use of resources.

Climate Resilience in Urban Planning

To prepare for climate change, cities need to take steps that reduce risks and make infrastructure stronger. SDG 11 focuses on several key areas:

- **Disaster Risk Reduction**: Cities need to have plans in place to deal with extreme weather events such as floods, droughts, and storms. As of **2022**, only about half of urban populations had access to such strategies. **Prague's Climate Change Adaptation Strategy** is an example of how a city can prepare for flooding through green infrastructure and better urban planning.
- **Green Infrastructure**: Expanding parks, planting more trees, and creating green spaces help reduce the heat trapped in cities due to concrete and buildings. Some cities are updating old buildings to make them more energy-efficient. Others are replacing outdated infrastructure with new solutions, such as wetlands that help prevent flooding. **Copenhagen** has set a goal to become carbon-neutral by **2025** through a mix of renewable energy and better transport planning.
- **Sustainable Transport**: Public transport systems help reduce traffic and lower emissions. SDG 11 aims to provide everyone with access to affordable and efficient transport by **2030**. Planning must consider the needs of all people, including women, children, and those with disabilities.
- **Inclusive Urban Development**: Cities need to be built in a way that benefits all communities. Better planning can help reduce inequality, improve access to housing, and strengthen public services.

- **Better Resource Management**: Cities need to reduce waste and use resources more efficiently. Recycling programmes, water conservation efforts, and energy-efficient buildings can make a difference in reducing overall waste and pollution.

Cities Leading Climate Resilience

- **Copenhagen**: The city is working to become carbon-neutral by **2025** by improving energy use, transport, and local policies.
- **Prague**: Green roofs and smart building management systems have been introduced to reduce energy use and help the city deal with rising temperatures.
- **Calgary & Edmonton**: These cities have changed zoning policies, upgraded infrastructure, and added wetlands in urban areas to prevent flooding.

Working Together for Climate-Resilient Cities

To prepare cities for climate change, different groups need to work together:

- **Governments** must create policies that support climate-friendly development.
- **Businesses** can invest in better technology and sustainable solutions.
- **Communities** can take part in local projects that improve their surroundings.

Better planning and action now can help cities become stronger and safer in the years to come.

SDG 12: Reducing Waste and Overproduction

As industries grow and people consume more, the amount of waste produced around the world keeps increasing. Resources are being used at a fast pace, and much of what is produced ends up discarded. Food is wasted, plastics fill the oceans, and many products are thrown away long before they have reached the end of their use. **Sustainable Development Goal 12 (SDG 12)** focuses on changing the way people and industries produce, use, and dispose of materials. It aims to reduce waste, manage resources responsibly, and create a system where economic growth does not come at the cost of the planet.

The Global Challenge of Waste and Overproduction

- **Resource Overuse**: The world's use of raw materials has increased rapidly. Between **2010 and 2017**, global domestic material consumption per person rose by **7%**, from **10.8 metric tonnes to 11.7 metric tonnes**. This growth reflects the increasing demand for materials such as metals, fossil fuels, and timber. If this pattern continues, natural resources will become even more strained in the coming years.

- **Food Waste**: Around **one-third of all food produced globally**, which amounts to more than **1.3 billion tonnes**, is wasted every year. This includes food lost during farming, processing, transport, and storage, as well as waste from supermarkets, restaurants, and homes. In some regions, such as **Latin America and the Caribbean**, over **500 kilocalories of food per person** are lost every day before even reaching consumers.
- **Plastic Pollution**: Plastics have become one of the most serious environmental challenges. Single-use plastics, such as straws, bags, and packaging, are a major part of the problem, as they often end up in landfills, rivers, and oceans. Some countries have introduced bans on certain plastic products, such as cotton buds, plastic cutlery, and polystyrene containers, to reduce pollution.

Targets Under SDG 12

- **Efficient Use of Resources (Target 12.2)**: By **2030**, the goal is to manage resources more efficiently and reduce unnecessary waste. This includes tracking the amount of materials used per person and ensuring that economic growth does not depend on increasing material consumption. In the **EU**, raw material consumption fell by **5% between 2018 and 2023**, showing progress towards using resources more carefully.

- **Reducing Food Waste (Target 12.3)**: The aim is to **halve global food waste per person by 2030** and reduce food loss throughout supply chains. Some of the ways this can be achieved include food recovery programmes, better storage methods, and improved transport systems.
- **Waste Management (Target 12.5)**: By **2030**, waste generation should be significantly reduced through prevention, reduction, recycling, and reuse. High-income countries currently recycle about **20% of their municipal solid waste**. In low-income countries, informal recycling workers collect and process large amounts of waste, but their efforts are often not officially recorded.
- **Sustainable Practices in Business (Target 12.6)**: Companies are encouraged to adopt sustainable practices and include environmental responsibility in their business plans. Some regions have introduced **Extended Producer Responsibility (EPR)** policies, which require manufacturers to take responsibility for the waste their products create. This includes covering the costs of recycling and disposal.
- **Public Awareness (Target 12.8)**: People need better access to information about sustainability so they can make more informed choices. Many consumers still do not have enough knowledge about how

their purchasing habits contribute to waste and environmental harm.

Reducing Waste and Overproduction

Circular Economy Models: Traditional production follows a **take-make-dispose** pattern, where products are used and then discarded. A **circular economy** focuses on keeping materials in use for as long as possible. This means designing products that can be repaired, reused, or recycled instead of being thrown away. In the **EU**, the circularity rate, which measures how much raw material is reused, was at **11.5% in 2022**. There is still room for improvement.

- **Technological Innovations**: New technologies are helping reduce waste. Advanced sorting systems for plastics, composting facilities for food waste, and material recovery processes are making recycling more efficient. Research is also being done to develop biodegradable plastics and alternative materials that break down naturally without causing pollution.
- **Policy Interventions**: Governments play a major role in reducing waste. Some have introduced bans on plastic products, while others provide financial incentives for businesses that use sustainable materials. Some governments have also removed subsidies that encourage overproduction, especially in industries like agriculture and fossil fuel extraction.

- **Behavioural Changes**: Individuals can help reduce waste by making simple changes. Reducing food waste at home, buying products with less packaging, and choosing reusable over disposable items can all make a difference. If millions of people make small adjustments, the overall impact can be significant.

Impact on Climate Change

Reducing waste and overproduction is directly linked to lowering greenhouse gas emissions. Every stage of production, extracting raw materials, manufacturing goods, transporting them, and disposing of waste, produces carbon emissions.

- **Food Waste**: Wasted food accounts for about **8–10% of global greenhouse gas emissions** every year. This is because food that decomposes in landfills releases **methane**, a greenhouse gas that is far more potent than carbon dioxide.
- **Recycling and Resource Recovery**: Recycling materials such as aluminium and paper requires far less energy than producing them from raw materials. This reduces emissions and lowers the demand for resource extraction, such as mining and deforestation.

Changing the way the world produces and consumes is essential to ensuring a more sustainable future.

SDG 13: The Direct Call for Urgent Climate Action

The effects of climate change are becoming clearer each year. Rising temperatures, more extreme weather, and natural disasters are affecting people and the environment across the world. **Sustainable Development Goal 13 (SDG 13)** calls for action to address these challenges, reduce the damage caused by climate change, and help communities adapt to its effects.

The Climate Crisis: A Global Emergency

- **Rising Temperatures**: Since **1880**, the Earth's temperature has increased by about **0.85°C**. If action is not taken, temperatures could rise by over **3°C** by the end of this century, far beyond the targets set in the **Paris Agreement**. The goal is to keep warming well below **2°C**, with efforts to limit it to **1.5°C**.
- **Extreme Weather Events**: Heatwaves, hurricanes, floods, and droughts have become more common and severe. These disasters destroy homes, affect food production, and cause financial losses. Communities that rely on farming and fishing are struggling with unpredictable weather and water shortages.
- **Communities Most at Risk**: Some of the poorest communities in the world are facing the worst effects of climate change. Coastal regions are experiencing rising sea levels,

and small islands are at risk of disappearing. Many people in these areas have limited resources to rebuild after disasters.

Targets of SDG 13

- **Building Stronger Communities (Target 13.1)**: Countries need to take steps to reduce the risks of climate-related disasters. This includes improving early warning systems, strengthening infrastructure, and protecting people from floods and extreme weather.
- **Including Climate Change in Policies (Target 13.2)**: Climate action needs to be part of national policies and long-term plans. Many governments have introduced **Nationally Determined Contributions (NDCs)** under the **Paris Agreement**, which outline their targets for reducing emissions.
- **Raising Awareness and Education (Target 13.3)**: People need access to better information about climate change and how to prepare for its effects. Education programmes and community initiatives can help people understand the changes happening around them and find solutions.
- **Funding Climate Action (Target 13.a)**: Developed countries have agreed to provide **$100 billion per year** to help developing nations fight climate change. This funding is meant to support projects that lower emissions and help vulnerable communities adapt.

- **Helping Vulnerable Countries (Target 13.b)**: Small island nations and developing countries need support to create climate action plans that suit their specific needs. Many of these regions contribute the least to emissions but face the greatest risks.

Urgent Actions Needed

- **Reducing Greenhouse Gas Emissions**: To prevent further warming, emissions must be cut by half by **2030**. This means moving away from fossil fuels and investing in clean energy like wind and solar power.
- **Adapting to Climate Change**: Countries need stronger infrastructure to deal with rising temperatures and extreme weather. Investments in flood protection, drought-resistant crops, and improved water systems can help communities cope with changing conditions.
- **Creating Green Jobs and Cleaner Cities**: The move to renewable energy and energy-efficient industries will create job opportunities and improve air quality. Making cities more sustainable by investing in cleaner transport and reducing waste will also help reduce emissions.
- **Ensuring Climate Finance is Delivered**: Developed nations need to meet their commitments under the **Paris Agreement** to support poorer countries in adapting to

climate change and reducing emissions. These funds can help countries build stronger infrastructure and switch to clean energy sources.

- **Working Together Across Borders**: Climate change affects everyone, and no country can solve it alone. Governments, businesses, and communities need to work together to develop solutions, share technology, and take responsibility for emissions.

Progress and Challenges

Many countries have introduced climate adaptation plans, but current pledges are not enough to keep temperatures within safe limits. **Global warming is on track to reach 2.7°C by the end of the century** if stronger actions are not taken.

The **IPCC Sixth Assessment Report** warns that even small increases in temperature will make climate risks worse. Every fraction of a degree adds to the damage caused by heatwaves, wildfires, and rising sea levels.

SDG 13 calls for urgent action to slow down climate change and protect future generations. If steps are not taken now, the effects will become even harder to reverse.

Reading about climate change and seeing its effects are two different things. It is one thing to know that temperatures are rising, but it feels different when you step outside in February and it already feels like summer. It is one thing to hear about extreme weather, but another to see floods destroy homes or watch wildfires turn forests into ash.

Change is happening all around us. Birds are migrating earlier. Cities are heating up. Crops are failing in places that once had steady seasons. These are not future problems. They are already here.

The world has plans. There are goals, agreements, and promises. Governments have pledged to cut emissions, industries are shifting to cleaner energy, and communities are finding ways to adapt. But looking at everything around us, the question is clear, **is this enough, or are we running out of time?**

Birds have always moved with the seasons, knowing when to stay and when to leave. Their world is shifting, and they are adjusting. The least we can do is make sure they still have a place to return to. **Let them fly, happily.**

Chapter 8:

E-Mobility – Driving the Future

The plan was simple, take a dip in the Ganga, visit Kashi Vishwanath and Kala Bhairava Swamy, and experience the spiritual energy of Varanasi. But like every journey, things did not go exactly as planned.

My wife and I were supposed to take an early morning **Air India** flight, but at **12:15 am**, we got a call from the airline informing us that the flight had been cancelled. We were about to leave home at **3:30 am** to catch the **7:00 am** flight. If my wife had not been awake when the call came, we would have gone to the airport only to find out the flight was cancelled.

I quickly booked another flight and secured our boarding passes. The change in plans meant a little more effort, but by **10:30 am**, we had reached **Varanasi** safely.

Everything started well. The **Kashi Vishwanath Temple** had been beautifully renovated. The new structure and improved facilities made it more organised for pilgrims. But as we moved through the city, I could not ignore the narrow roads and the slow-moving traffic. Most people were getting around in small **rickshaws** and **bicycle taxis**. The movement of vehicles in the central part of the city was painfully slow, just like in big cities like **Bangalore** and **Mumbai**, where traffic takes over daily life.

The slow movement of vehicles in crowded cities increases pollution, especially **PM 2.5**, which affects health in many ways. With cities growing rapidly, the problem is only getting worse.

By **2060**, the world's population is expected to reach **9 billion**, and **70%** of people will be living in cities. That means around **6.3 billion people** will be packed into just **4%** of the Earth's land. This is **1.5 times the current global population**.

With so many people in cities, the demand for **tall buildings, highways, metros, railways, bridges, and tunnels** will increase. The number of roads, vehicles, and transport systems will grow at least **three times** within this small portion of land.

The question is, **how do we build cities that can handle this growth without making life more difficult?**

One of the biggest challenges in cities is **transportation**. The way people move from one place to another affects daily life and the environment. Roads are getting more crowded, travel times are increasing, and pollution levels are rising. The more vehicles we put on the roads, the more emissions we release into the air.

Transport is a major contributor to **carbon emissions**, with cars, buses, and trucks running on petrol and diesel adding huge amounts of pollution every day. In cities with slow-moving traffic, emissions are even worse because vehicles burn fuel while stuck in jams. The more time people spend on the road, the more harmful gases enter the air.

Many cities around the world are already struggling with this. The air in places like **Delhi, Beijing, and Los Angeles** is filled with pollutants, making it difficult to breathe. Health problems linked to air pollution, such as asthma and lung diseases, are increasing. If cities continue to grow without better transport systems, this problem will only get worse.

The way we travel needs to change. Cities need **better public transport**, more efficient roads, and cleaner ways to move. Investing in **metro systems, electric buses, bike lanes, and pedestrian-friendly streets** can help reduce emissions and improve life for millions of people. Some countries have already started working on this, but the changes are happening too slowly compared to how fast cities are expanding.

*With more people moving to urban areas, the question is, **how can transportation keep up without making pollution worse?***

With more people moving to cities, the way we travel is becoming a bigger challenge. Roads are crowded, traffic moves slowly, and pollution levels keep rising. Public transport is often unreliable, and many people still depend on petrol and diesel vehicles. This is why **sustainable mobility** is so important. It focuses on finding cleaner, smarter ways to move without harming the environment.

What is Sustainable Mobility?

Sustainable mobility means using transport systems that reduce pollution, save energy, and make cities more liveable. This includes **electric vehicles, better public transport, cycling, walking, and shared transport options**. The goal is to cut down emissions while making travel safer, easier, and more efficient.

Transport is one of the biggest contributors to air pollution and carbon emissions. In cities, vehicles release harmful gases that make the air unhealthy to breathe. Poorly planned roads and outdated public transport make the problem worse. If cities continue to expand without better mobility options, pollution and traffic congestion will increase, affecting millions of lives.

Governments and city planners play a important role in making transport cleaner and more efficient. Policies that **promote electric vehicles, improve public transport, and support cycling and walking** can make a big difference. Many countries are offering subsidies for EV buyers, adding more charging stations, and introducing fuel efficiency standards for vehicles.

Communities also have a role to play. People can choose more sustainable ways to travel, whether by using public transport, cycling, or carpooling. Businesses can support employees by offering transport benefits that encourage shared rides or remote work options.

Sustainable transport is about **reducing pollution, making travel easier, and improving the quality of life** in cities. This means using **cleaner energy sources, better public transport, and smarter city planning** to reduce the number of vehicles on the road.

Why It Matters

- **Better Air Quality** – Cities with high traffic often have some of the worst air pollution levels. Switching to cleaner transport options can help reduce lung diseases and other health problems caused by poor air quality.
- **Lower Emissions** – The transport sector is one of the biggest contributors to **carbon emissions**. Sustainable transport options like **electric buses, trains, and shared mobility services** can help reduce emissions and slow down climate change.
- **More Efficient Cities** – Well-planned transport systems can save time and money. Faster and more reliable public transport, better cycling lanes, and pedestrian-friendly streets can make cities easier to navigate without depending on cars.
- **Cost Savings** – Cleaner transport options **reduce fuel costs**, lower maintenance expenses, and decrease healthcare costs linked to pollution-related diseases.

But the solution?

Electric Vehicles (EVs)?

Electric vehicles (EVs) are becoming more common on our roads. Unlike traditional petrol or diesel cars, EVs run on electricity and do not produce exhaust emissions. This change is important for reducing air pollution and improving public health.

Traditional vehicles emit pollutants like nitrogen oxides and particulate matter, which can cause health problems such as asthma and heart disease. EVs, on the other hand, have no tailpipe emissions, leading to cleaner air in cities. A study by the European Environment Agency found that, over their lifetime, EVs produce fewer greenhouse gases and air pollutants than conventional cars.

While producing electricity for EVs can generate emissions, these are usually lower than those from traditional vehicles. As more renewable energy sources like wind and solar are used, the environmental benefits of EVs increase. The European Environment Agency reports that, even with current energy mixes, EVs contribute to lower overall emissions.

Cities with high traffic often suffer from poor air quality. EVs can help address this issue by reducing local air pollution. Research indicates that increased use of EVs leads to significant improvements in urban air quality, benefiting all residents.

Despite their benefits, EVs face challenges. The production of EV batteries can result in higher initial emissions, but these are offset over time by the vehicle's lower operational emissions. Additionally, the environmental impact of EVs depends on how the electricity they use is generated. As the energy grid becomes greener, the advantages of EVs will continue to grow.

Green Public Transport?

Travelling should not feel like a daily struggle, but in many places, that is the reality. **Public transport powered by clean energy can help solve this problem by reducing pollution and making travel easier.**

Buses carry millions of people every day, but many still run on diesel, releasing smoke and harmful gases. Some cities are replacing them with **electric buses**, which run without producing pollution. In **Manchester, 330 new electric buses** have been introduced as part of the **Bee Network**, improving public transport while keeping the air cleaner.

Electric buses move smoothly and do not release harmful fumes. More charging stations and better batteries are being developed so these buses can run longer without needing frequent stops. Many countries are investing in electric buses as part of their plans to reduce pollution.

Trains move large numbers of people quickly and efficiently, but many still use diesel, which adds to pollution. **Electric and hydrogen-powered trains** are providing cleaner alternatives.

In **California**, the **Zero-Emission Multiple Unit (ZEMU)** has been introduced as **North America's**

first hydrogen-powered passenger train. It runs without polluting the air, showing how railway transport can be improved with clean energy. Many countries are testing similar trains to cut emissions and modernise their transport networks.

Not every trip requires a car. Many cities now have **bike rentals, e-scooters, and car-sharing services**, helping people get around without adding more cars to the road. In **North America**, shared transport options have **prevented 74 million pounds of CO_2 emissions**, showing how small changes in travel choices can make a big difference.

Carpooling is another way to reduce pollution. Some companies encourage employees to share rides, and public transport systems are working to connect people with ride-sharing services. Fewer cars on the road mean less congestion and lower emissions.

Governments are investing in **better bus services, metro systems, and cycling routes** so that people have more choices beyond driving. Countries like **Germany** are funding cleaner fuels, including **electricity-based fuels and biofuels**, to improve public transport while lowering emissions.

While cities are adding more electric buses and cleaner trains, India is also taking a big step in upgrading its rail systems. A major high-speed train project is now under way, connecting Mumbai and Ahmedabad. This could change how people travel across the country.

India's High-Speed Rail Project

India is building its first bullet train between Mumbai and Ahmedabad. This route is called the Mumbai–Ahmedabad High-Speed Rail Corridor. It will cover about **508 kilometres**, and the trains are expected to run at speeds of up to **320 kilometres per hour**. With this, the travel time between the two cities will come down from nearly **seven hours** to around **two hours**.

The project is being handled by the **National High Speed Rail Corporation Limited (NHSRCL)**. It is using Japanese technology known as the **Shinkansen system**, which is famous for being fast, safe, and very reliable.

The first part of the train line that will be ready is between **Surat and Bilimora** in Gujarat. This stretch will be used first for trial runs and later for regular service. The full line from Mumbai to Ahmedabad is expected to be completed by **2030**, though some parts may be ready earlier.

The main idea behind this project is to give people a faster and cleaner way to travel. Since these trains run on electricity, they are better for the environment compared to diesel trains or flights. This project will also help improve the country's transport system and connect major cities more efficiently.

There are bigger plans ahead. A study has suggested building around **10,000 kilometres** of high-speed rail across India by the year **2035**. This plan includes new train routes connecting cities like **Delhi, Kolkata, Chennai**, and many others. If this happens, people will be able to travel across the country much faster and more easily.

While fast trains are changing how we move across cities and states, cars are going through their own transformation. One of the biggest challenges for electric vehicles has been charging time. Waiting for hours to charge a car has been a problem for many people thinking about switching to electric. That is where something new from China caught my attention.

BYD's Fast-Charging Technology

A company called **BYD**, based in China, has introduced something that could change how we look at electric vehicles. They have developed a new **ultra-fast charging battery** that can fully charge an electric car in **just five minutes**. This technology was showcased in early 2024 and has already started conversations around the world.

The new battery is called the **Blade Battery**, and BYD has been working on improving both its performance and safety. Their latest version, paired with a fast-charging system, is able to power up a car in the same time it takes to fill petrol at a regular station. That changes things.

To make this possible, BYD has upgraded the battery's structure and added a new cooling system that prevents overheating. The company has also built new charging stations that support this level of fast charging. These stations deliver very high electricity power safely and quickly, which helps in cutting down waiting time.

This development is more than just a technical step. It speaks to a bigger shift in how we think about electric vehicles. If charging becomes faster, more people might be willing to choose electric cars. It means no long waits, less worry, and more freedom on the road. For people living in cities or those who travel often, this can be a big improvement.

The five-minute charge also helps in reducing traffic at charging points. If more cars can charge and leave in less time, then queues can be avoided. This makes life easier for both drivers and city planners.

While BYD is setting new standards in fast charging, another big name in the electric vehicle world is getting ready to make its move in India. After years of talks and speculation, it looks like Tesla is finally coming in.

Tesla's Entry into India

In early 2024, news broke that **Tesla** had officially registered a company in India under the name **Tesla India Motors and Energy Pvt Ltd**. The announcement was followed by reports of meetings between Tesla executives and government officials to discuss setting up production and sales in the country.

Tesla is expected to begin its India journey with **Model 3** and **Model Y**, which are its more affordable models. These cars will likely be introduced either as fully imported units to start with, or assembled locally if agreements around production are made. The government has shown interest in working with Tesla, especially if it involves building a **manufacturing plant** in India in the future.

The arrival of Tesla could shake up the electric car market in India. Prices of premium EVs may start becoming more competitive. Local car makers like **Tata Motors**, **Mahindra**, and even international players such as **Hyundai** and **BYD** will have to rethink how they price and design their electric cars.

Recently JSW Group led JSW-MG Motors India company has quietly reshaped the EV landscape with the introduction of its Battery-as-a-Service (BaaS) model, starting with the MG Windsor EV with a BaaS plan has a lower upfront purchase price and a monthly fee based on kilometers driven.

Instead of owning the battery outright, customers can now lease or subscribe to it — paying only for the energy they use. It's a simple shift, but one with huge implications for how we buy, use, and think about electric cars.

Why It Matters

Lower upfront cost: The battery makes up nearly 40% of an EV's price. Removing it from the sticker cost makes EVs more affordable to the masses.

Faster adoption: A pay-as-you-go model reduces financial barriers and accelerates EV penetration in cost-sensitive markets.

Easy upgrades: Drivers can upgrade to newer battery tech or swap out depleted packs in minutes.

Ideal for fleets: Logistics, ride-share, and taxi operators can keep vehicles running continuously — no downtime for charging.

Sustainability built-in: MG retains responsibility for battery health, recycling, and second-life applications.

Beyond Ownership — Toward Mobility as a Service

With this move, MG isn't just selling a car; it's selling mobility freedom.

The "Battery-as-a-Service" approach could redefine the relationship between drivers, vehicles, and energy — just as subscription models transformed entertainment and software.

It has become very successful and has set a new standard for India's EV ecosystem, and maybe even globally how the Electric vehicles are manufactured and sold.

Apart from cars, Tesla's entry could help improve the charging infrastructure. Tesla brings its own **Supercharger network**, which has helped build confidence in electric vehicles in other countries. If the same is introduced in India, it could push other companies to improve their own charging networks.

Tesla's brand name also carries weight. Many people know it as the company that made electric cars popular across the world. Its presence in India could encourage more people to try electric cars, especially younger drivers in cities.

All of this shows how quickly the EV space in India is changing. New players, better batteries, and faster charging are all coming together. The road ahead looks very different from what it was a few years ago.

All of this made me wonder, if our cars are changing this fast, what about the way we travel beyond the roads? Trains are picking up speed, cars are going electric, and cities are slowly getting ready for a cleaner future. But what happens when we start thinking beyond the ground?

From Air Travel to Space Journeys?

As our world grows, so does our need to explore and connect. From the planes we board for cross-country trips to the ambitious plans of reaching other planets, transportation is evolving rapidly.

Air Travel

Airplanes have changed how we experience the world, making distant places accessible in hours. Modern aircraft are becoming more efficient, aiming to reduce fuel use and emissions. The aviation industry is exploring sustainable fuels and electric engines to make flying kinder to our planet.

Sea Travel

Ships play a crucial role in global trade, moving goods across oceans. Efforts are underway to make sea travel greener, such as developing vessels that produce fewer emissions. While sea transport is slower than air travel, it is more cost-effective for transporting large quantities of goods.

High-Speed Trains

High-speed trains, like Japan's Shinkansen and France's TGV, have transformed land travel by connecting cities swiftly. These trains can reach speeds over 300 km/h (186 mph), cutting down travel times significantly. Plans are in place to expand these networks, including proposals for underwater routes that could connect distant regions more directly.

Water Transport

Beyond the open seas, rivers and canals serve as vital pathways for moving people and goods. Innovations in this area include cleaner engines and modernized ports to handle more traffic efficiently.

Interplanetary Travel and SpaceX's Starship

Looking beyond Earth, ambitious projects aim to make space travel a reality for more than just astronauts. SpaceX, founded by Elon Musk, is leading with its Starship spacecraft. Designed for missions to the Moon, Mars, and beyond, Starship is a fully reusable spacecraft intended to carry both crew and cargo. Musk envisions launching uncrewed missions to Mars within the next two years to test landing capabilities, with crewed missions potentially following in about four years. The ultimate goal is to establish a self-sustaining city on Mars within 20 years, making humanity a multiplanetary species.

Infrastructure?

Electric vehicles are becoming more common, but the challenge is keeping them charged and ready to use. Without enough charging stations, many people hesitate to switch from petrol or diesel cars. Governments and private companies are working on building more charging points in cities, along highways, and in residential areas to make sure people can charge their vehicles without difficulty. The **United States** has a plan to set up **500,000 public charging stations by 2030**, while **India** and **European countries** are expanding their charging networks to support the growing number of electric vehicles on the road.

Battery technology is improving to help vehicles run longer on a single charge. One of the biggest issues has been the time it takes to charge a car. Many companies are developing faster-charging batteries. **China's CATL** has introduced batteries that can provide a **600-kilometre range after just 10 minutes of charging**. Research is also being done to create batteries that last longer, so cars can travel greater distances without needing to stop for a charge.

Karnataka, India, has launched a **Clean Air Action Plan**, which includes measures to encourage people to use electric vehicles. Reducing pollution from petrol and diesel cars is one of the ways cities are trying to improve air quality. More electric buses, taxis, and two-wheelers are being introduced to reduce emissions in crowded areas.

Charging electric vehicles with clean energy is just as important as building more charging stations. If the electricity used for charging comes from coal or other fossil fuels, the environmental benefits are reduced. Countries are investing in wind, solar, and other renewable energy sources to make sure electric vehicles are powered in a way that does not add to pollution. **Sweden** is developing **hydrogen-based charging stations**, which provide an alternative to traditional electricity-powered chargers and make use of cleaner energy sources.

The Shift to Green Transport is Already Happening

The way we travel is changing. Roads are no longer filled only with petrol and diesel cars. More electric vehicles are appearing, quietly moving through traffic. Charging stations are being built in parking lots, fuel stations, and street corners. Buses that once released dark fumes are being replaced with cleaner electric models. It no longer feels like something that belongs to the future, it is happening now.

More people are choosing electric vehicles because they need less maintenance and cost less to run in the long term. Governments are making them more affordable with tax benefits and incentives. Battery technology is improving, helping cars travel longer distances before needing a charge. Public transport is also changing. Many cities are introducing electric buses, and shared mobility services like bike rentals and car-sharing platforms are becoming more common.

The transition has begun, but challenges remain. Many countries are working on expanding charging networks and making electric transport more accessible. ***The question is no longer whether green transport will replace traditional vehicles. It is happening. The real question is, how fast will this change take place, and will it be enough?***

Chapter 9:

The Energy Transition – Breaking Free from Fossil Fuels

It had been a long day. The kind where everything feels slow, where even after leaving work, the heat follows. The streets were crowded, the buses full, the traffic moving in short bursts. The sun had been harsh all afternoon, and by the time I got home, I wanted nothing more than to sit down and cool off.

I stepped inside, and the power went out.

The fan stopped, the lights went off, the room felt smaller. It was still warm from the day, and without any way to cool down, the heat settled in. I waited, thinking it would come back in a few minutes. It did not.

I stepped outside. The whole street was dark. A few neighbours were standing near their gates, looking around, waiting for something to change. Some sat on their porches, waving their hands in the heat. Someone said it could take a while.

It took hours.

By the time the power returned, it was late. The lights flickered back, the fan started again, everything went back to how it was. But as I sat there, I kept thinking about it. How everything depends on something we cannot see. The lights, the transport, the machines that keep stores running, the factories that produce the things we buy. Every part of life needs it. And when it stops, we realise how much we rely on it.

The power cut had ended, but the thought remained. ***How do we finally break free from fossil fuels?***

Our Dependence on Fossil Fuels

Fossil fuels, coal, oil, and natural gas, have been the main sources of energy for the world since the Industrial Revolution. They power our homes, vehicles, industries, and much more. In 2023, fossil fuels made up about 80% of the world's primary energy consumption. This heavy reliance has several consequences:

Environmental Impact

Burning fossil fuels like coal, oil, and natural gas releases large amounts of carbon dioxide (CO_2) into the atmosphere. This increase in CO_2 levels leads to global warming and climate change.

In 2023, global CO_2 emissions from fossil fuel combustion and industrial processes reached 37.01 billion metric tonnes. This was a 1.08% increase from the previous year, setting a new record high.

Since 1990, these emissions have risen by more than 60%.

Energy production and use are the largest sources of human-caused greenhouse gas emissions.

The International Energy Agency (IEA) reports that energy-related activities account for over three-quarters of global greenhouse gas emissions.

In the United States, CO_2 emissions made up about 80% of total human-made greenhouse gas emissions in 2022. Burning fossil fuels was responsible for 74% of total greenhouse gas emissions and 93% of all human-made CO_2 emissions in the country.

The continuous burning of fossil fuels has significantly increased atmospheric CO_2 concentrations. As of May 2022, CO_2 levels were nearly 50% higher than pre-industrial times.

This rise in CO_2 levels intensifies the greenhouse effect, leading to higher global temperatures and contributing to climate change.

Economic Concerns

The global economy relies heavily on fossil fuels like oil, natural gas, and coal. The availability and pricing of these resources can be unpredictable, often influenced by geopolitical events. This unpredictability can lead to economic instability, affecting household energy bills and the cost of goods and services.

Geopolitical events can cause significant fluctuations in fossil fuel prices. For example, the Russian invasion of Ukraine in 2022 led to a sharp increase in oil and gas prices. The conflict disrupted global energy markets, causing a surge in prices and contributing to higher living costs worldwide.

Trade disputes and tariffs can impact fossil fuel prices. In 2025, under President Donald Trump's administration, new tariffs led to increased costs for imported goods, including energy resources. This resulted in higher prices for consumers and added to inflationary pressures.

Fluctuations in fossil fuel prices directly affect household energy bills. During periods of high oil prices, consumers may face increased costs for heating, electricity, and transportation. Conversely, sudden drops in prices can lead to economic instability in regions dependent on oil production, affecting employment and public services.

The unpredictability of fossil fuel prices can influence the cost of goods and services. As transportation and production costs rise with increasing fuel prices, businesses may pass these expenses onto consumers, leading to higher prices for everyday items.

Health Issues

Burning fossil fuels like coal, oil, and natural gas releases pollutants into the air, leading to serious health problems. The World Health Organization (WHO) reports that exposure to air pollution causes about 7 million premature deaths each year.

Air pollution from fossil fuels contributes to various health issues:

- **Respiratory Diseases**: Pollutants can cause asthma, reduced lung function, and infections.

- **Cardiovascular Diseases**: Exposure increases the risk of heart disease and strokes.
- **Cancer**: Combustion of fossil fuels releases harmful particles and hydrocarbons, increasing cancer risk.

The Hidden Danger

Carbon dioxide is everywhere. It is invisible, but its impact is felt in every corner of the world. It is not like smoke from a factory chimney or the dust that settles on city streets. It does not leave a trace that we can see, but it lingers, trapping heat, changing the climate, and affecting the air we breathe.

When fossil fuels burn, they release CO_2 into the atmosphere. Unlike other pollutants that eventually settle or wash away, CO_2 stays for centuries. The more it builds up, the worse the problem becomes. Temperatures rise, weather patterns shift, and the world changes in ways that affect everything from the food we eat to the air we inhale.

The effects of CO_2 are far-reaching. It makes extreme weather more common, leading to stronger storms, longer droughts, and unpredictable seasons. It dissolves into the ocean, making the water more acidic and harming marine life. It disrupts plant growth, reducing nutrients in crops and affecting food security. It even changes the balance of oxygen in the air, lowering air quality in ways that go beyond dust and smoke.

The World Health Organization has warned that air pollution, largely driven by CO_2 emissions, is responsible for millions of premature deaths each year. The longer this continues, the worse it gets. Cutting down CO_2 emissions is not just about preventing climate change; it is about survival. Every step taken to reduce CO_2 ensures cleaner air, a stable climate, and a healthier future.

The damage caused by CO_2 is clear. The question is, **how much longer can we ignore it?**

Transitioning to renewable energy sources is essential for a cleaner and more sustainable energy grid. Solar, wind, hydro, and geothermal energy each play a unique role in reducing our reliance on fossil fuels and decreasing greenhouse gas emissions.

Solar Energy

Solar energy involves capturing sunlight and converting it into electricity using solar panels. This method provides a clean and plentiful source of power.

In 2024, the United Kingdom achieved its cleanest electricity generation on record, with 58% coming from low-carbon sources, including record outputs from wind and solar. Renewable energy sources, such as wind, solar, and hydropower, are projected to generate more electricity in the UK than fossil fuels for the first time, according to the think-tank Ember. Renewables are expected to account for around 37% of the electricity output, compared to 35% from fossil fuels. This shift has significantly reduced carbon emissions and reliance on fossil fuels.

The growth of solar energy in the UK has been remarkable. In 2023, solar power contributed 4.5% to the UK's renewable energy sources. This increase reflects the country's commitment to expanding its renewable energy capacity and reducing its carbon footprint.

The expansion of solar energy has provided environmental benefits and economic advantages. The UK's net zero economy is expanding significantly, growing three times faster than the general economy, according to the Confederation of British Industry (CBI). In 2024, the net zero sector grew by 10% and contributed £83 billion in gross value added (GVA). This growth underscores the positive impact of renewable energy investments on the nation's economy.

Wind Energy

Wind energy uses turbines to convert wind into electricity. It is a rapidly growing renewable energy source worldwide.

In 2023, the global wind industry installed a record 117 gigawatts (GW) of new capacity, marking a 50% increase from the previous year. This growth highlights wind energy's expanding role in the global energy landscape.

China leads in wind energy capacity, accounting for over 50% of the world's installed wind capacity as of 2023. The United States and Germany also contribute significantly to global wind energy production.

Wind energy offers several benefits, including reducing greenhouse gas emissions and promoting energy security. It also stimulates economic growth by creating jobs in manufacturing, installation, and maintenance of wind turbines.

Hydropower

Hydropower is one of the oldest and most widely used renewable energy sources in the world. It generates electricity by using the movement of water, usually from rivers or dams, to spin turbines that produce power. Unlike fossil fuels, hydropower does not release carbon dioxide or other harmful pollutants into the air, making it one of the cleanest ways to produce electricity.

Hydropower provides about **16% of the world's electricity** and remains the largest source of renewable energy. Countries like **China, Brazil, Canada, and the United States** are among the biggest producers, with China leading the way. In 2023, China generated more than **1,300 terawatt-hours** of electricity from hydropower, meeting a large portion of its energy needs. Brazil also relies on hydropower for nearly **60% of its electricity**, using large dams like the **Itaipu Dam**, which is one of the largest hydropower plants in the world.

Hydropower plays a major role in **energy security** because it provides a steady supply of electricity. Unlike solar or wind power, which depend on the weather, hydropower plants can produce energy continuously. Many countries use hydropower for **baseload electricity**, meaning it supplies a constant level of power to the grid. In addition, hydropower plants can quickly adjust their output, making them useful for balancing energy demand and supply.

Beyond electricity, hydropower also supports **water management**. Dams store water, helping with irrigation, flood control, and drinking water supplies. Some large hydropower projects have transformed entire regions by providing both energy and a reliable water source for agriculture and daily use.

Geothermal Energy

Geothermal energy taps into the Earth's internal heat to produce electricity or provide direct heating. This method offers a stable and continuous energy supply, as the Earth's heat is always available, unlike solar or wind energy, which depend on weather conditions.

Currently, geothermal energy contributes a small portion to the global energy mix. As of 2024, it accounts for about 0.4% of worldwide electricity generation, with a total installed capacity of around 16 gigawatts (GW). However, its potential is vast, especially in regions rich in geothermal resources. For instance, the United States has the largest cumulative geothermal power capacity installed but is still using less than 1% of its total geothermal resources. Kenya produces the largest share of its electricity from geothermal (46%), while the Philippines, Chile, and Turkey have been adding significant capacity in recent years.

Enhanced Geothermal Systems (EGS) are being developed to unlock this potential. Unlike traditional geothermal systems that require specific conditions like natural hot water reservoirs, EGS can be implemented in a wider range of locations. This is achieved by creating artificial reservoirs in hot rock through techniques such as hydraulic stimulation, allowing water to circulate and capture heat. The U.S. Department of Energy estimates that EGS could power over 65 million American homes and businesses in the future.

Recent technological advancements have made EGS more viable. For example, in 2023, Fervo Energy successfully demonstrated a pilot EGS plant in Nevada, generating 3.5 megawatts (MW) of continuous power. This project showcased the commercial potential of EGS and opened the door for larger-scale developments.

The International Energy Agency (IEA) suggests that with continued innovation and investment, geothermal energy could become cost-competitive with wind and solar power within the next decade.

Challenges in Adopting Renewable Energy

Transitioning to renewable energy is essential for a sustainable future, but it comes with several challenges:

Cost

While the costs of renewable energy technologies like solar and wind have decreased significantly, the initial investment remains substantial. For example, the U.S. Department of Energy estimates that upgrading electricity distribution networks to accommodate renewables will require between £37 billion and £50 billion by 2050. This investment could add between £5 and £25 annually to household costs until 2050. However, delaying these investments may lead to higher long-term expenses.

Infrastructure Gaps

Integrating renewable energy into existing power grids demands significant infrastructure upgrades. In the UK, plans to decarbonize electricity by 2030 involve installing new transmission lines, which has faced opposition from local communities concerned about the impact on landscapes and tourism. Balancing the need for new infrastructure with community interests presents a complex challenge.

Political Hurdles

Political factors can influence the adoption of renewable energy. In Germany, upcoming elections have shifted focus from climate policies to issues like immigration and the economy. Leading candidates have indicated a preference for economic strength over strict climate initiatives, potentially impacting the country's commitment to renewable energy and influencing global climate action efforts.

Investing in Renewable Energy for a Sustainable Future

Energy is at the centre of everything. It keeps homes running, powers industries, and drives economies. But the way the world produces energy has long depended on fossil fuels, bringing pollution and instability. With climate change worsening and energy security becoming a growing concern, countries are putting more focus on renewables.

In 2023, global investments in clean energy reached **$1.77 trillion**, showing a strong shift towards sustainable solutions. Countries are recognising that relying on fossil fuels comes with risks, unpredictable prices, supply shortages, and environmental damage. The United Kingdom's **net zero economy** grew by **10% in 2024**, contributing £83 billion and supporting nearly a million jobs. Renewable energy is not just about the environment. It is creating industries, jobs, and economic growth.

There are challenges. Some companies have slowed down investments in renewables because of concerns about costs. National Grid, for example, sold its **U.S. onshore renewables business** to focus on its main operations. But despite these concerns, investors see long-term potential. Experts believe that shifting to clean energy is not just necessary, it makes financial sense. Governments and private investors are looking at renewables as a strong and reliable investment for the future.

Efforts are also being made to expand renewables in developing countries. The UK's **British International Investment** is financing renewable projects in emerging markets, helping regions that struggle with energy access. A cleaner, more stable energy system benefits everyone.

The night had settled in. The streets were dim, and the power had gone again. A neighbour stepped outside, looking at the darkness, waiting for the electricity to return. The houses stood still, waiting for the hum of machines to start again. This was not a rare event. It was a reminder.

How much longer will we wait for a real change?

Chapter 10:

Net Zero – A Personal Vision

It started with a conversation.

I had met an old friend after years. We sat in a café that had changed owners too many times, but the tables and chairs were the same. He had always talked about things ahead of their time, new ideas, inventions, the kind of things most people never thought about.

This time, he spoke about space.

"Do you know they're planning power stations outside Earth?" he said.

I thought he was joking. He wasn't.

"Solar stations in orbit. No clouds, no nights, just endless energy. It's already being tested."

I waited for him to explain. He leaned forward.

"The sun is always shining in space. If we build stations up there, they can send energy back to Earth using microwaves. No more power cuts, no more fuel shortages, just constant electricity."

I listened, trying to picture it. The way he spoke, it felt like it was already happening.

"Japan has been working on this for years. China wants a working station in orbit before

2035. Private companies are investing in it. It's not just an idea anymore."

I looked at him, thinking about everything we had once thought was impossible.

"We used to say the same about electric cars," I said.

He nodded. ***"And solar panels. And wind power. Now, look around. They are everywhere."***

It made sense. The world had been moving forward, step by step. Change never felt fast, but when you looked back, everything was different.

As I left, I kept thinking about it. How life had been built around things that could run out. How people had already started working on what came next.

The future was not waiting for permission. It was being built. **And that made me think about something bigger, Net Zero, my personal vision.**

For years, I had read about climate change, about how the world was heating up, about the disasters that were becoming more common. I had seen reports of rising emissions, power shortages, fuel prices going up. But those were just headlines, things that seemed far away. It wasn't until I looked around, really looked, that it felt real.

I thought about the long summers that felt hotter than before. The days when the power went out and everything stopped. The traffic that never seemed to move, the fuel prices that always seemed to rise, the air that felt heavier in cities filled with cars and factories. It was everywhere. It was daily life.

And that was when I realised, Net Zero was not just some distant goal set by governments or businesses. It was a necessity. It was not something that would happen on its own. It was something that needed action, from countries, from industries, from people.

I thought about everything that had been discussed, the changes that were already happening. Solar power, wind energy, electric transport, better ways to build cities. The ideas were there. The solutions were there. The question was, *how fast could they become the norm?*

Net Zero wasn't about reaching a perfect world. It was about making sure the world didn't get worse. It was about cutting down emissions, using cleaner energy, changing how things worked so that future generations wouldn't have to fix the damage left behind. It was about making choices, every day, in small and big ways, that helped move things forward.

The more I thought about it, the more it became clear. The world would either move towards Net Zero by design or by disaster. The choice was still there. **But for how long?**

This was about restoring balance. Some emissions would always exist, but they had to be removed. Energy would still be needed, but it had to come from sources that did not make things worse. Moving towards Net Zero was not one answer. It was a system that had to work together.

And that led me to my own vision and actions.

Moving Towards Net Zero Emissions

Reaching net zero emissions means balancing the amount of greenhouse gases released into the atmosphere with the amount removed. Some emissions can be reduced through cleaner energy sources, better efficiency, and sustainable industry practices. However, certain activities, such as agriculture and manufacturing, will continue to produce emissions that cannot be avoided. These must be removed to prevent further damage to the climate.

One way to reduce emissions is by switching to renewable energy. Solar, wind, and hydroelectric power do not produce carbon dioxide when generating electricity. Public transport, electric vehicles, and improved infrastructure can also lower emissions by cutting reliance on petrol and diesel. Energy-efficient buildings and industrial processes can further reduce waste and pollution.

Even with these changes, some emissions will still remain. This is where carbon removal becomes necessary. Trees absorb carbon dioxide as they grow, making reforestation one of the simplest ways to remove emissions. Restoring soil health can also help absorb carbon naturally. Advanced technologies, such as direct air capture, pull carbon dioxide from the atmosphere and store it underground or use it for industrial applications.

Reducing emissions and removing existing pollution are both needed to stabilise the climate. It is not about eliminating everything at once but ensuring that the world does not continue adding to the problem. Investing in both reduction and removal will determine how soon net zero can become a reality.

Net Zero Buildings: Energy-Efficient Designs and Sustainable Materials

Buildings use a large amount of energy. They need power for heating, cooling, lighting, and running appliances. In 2018, almost **40% of the world's carbon emissions** came from constructing and operating buildings. This is why there is a growing effort to design buildings that use less energy and create less pollution.

Energy-Efficient Designs

The way a building is designed affects how much energy it needs. Some buildings stay cool without air conditioning, while others trap warmth in winter, reducing the need for heating.

- **Better Insulation and Airtight Structures**: Thick insulation and airtight designs help keep buildings warm in winter and cool in summer. A school in Brooklyn, New York, was built using these methods and now uses far less energy than other schools in the city.
- **Natural Light and Ventilation**: Using large windows and open spaces allows fresh air to move through buildings and reduces the need for artificial lighting. A home in Oxford was designed with this in mind, helping

lower energy use while keeping the space comfortable.

- **Efficient Heating and Cooling Systems**: Heating, ventilation, and air conditioning (HVAC) systems use a lot of energy. Newer buildings are using **geothermal systems,** which use underground temperatures to heat and cool spaces without burning fuel.

Sustainable Materials

The materials used in buildings can also help lower emissions. Some materials require a lot of energy to produce, while others are more environmentally friendly.

- **Recycled and Reclaimed Materials**: Using wood, metal, or stone from old buildings reduces waste and the need for new materials. A home in Oxford used recycled copper for its exterior, making it both durable and sustainable.
- **Low-Carbon Materials**: Wood from responsibly managed forests, bamboo, and materials like **hempcrete** are becoming more popular because they are better for the environment than concrete and steel.
- **Locally Sourced Materials**: Transporting materials from far away adds to pollution. Sourcing wood, stone, and other materials from nearby areas reduces emissions and supports local businesses.

Examples of Net Zero Buildings

Some buildings already show what is possible with better design and materials.

- **Pearl River Tower in Guangzhou**: This skyscraper has wind turbines, solar panels, and a special outer layer that keeps the temperature inside stable, reducing the need for heating and cooling.
- **Millfield Green Retirement Village in Luton**: This community was built with solar panels, ground-source heat pumps, and extra insulation, lowering energy costs for residents while reducing emissions.

By designing buildings to use less energy and choosing better materials, cities can cut emissions and make homes, offices, and public spaces more efficient.

E-Mobility: The Importance of EVs and Green Public Transport

Electric vehicles (EVs) and cleaner public transport are changing the way people travel. More countries are moving towards these options as pollution from petrol and diesel vehicles affects air quality. Governments and industries are working to make these changes happen faster by improving technology and making clean transport more accessible.

More people are choosing electric cars. In 2024, over 15 million EVs were sold worldwide, which was more than the year before. In China, every second car sold was electric. In Norway, almost all new cars were electric.

Transport causes about 14% of the world's carbon emissions. EVs help reduce pollution because they do not release smoke or harmful gases. This makes the air cleaner, which is better for people's health. Many countries are encouraging people to buy EVs by offering tax benefits and subsidies. Charging stations are also increasing, so people do not have to worry about where to charge their cars.

Battery technology is improving. Some companies have made batteries that can charge in ten minutes and allow cars to travel up to 600 kilometres. These improvements make electric cars more convenient for daily use.

Buses, trains, and other public transport systems are becoming cleaner. Many cities have started using electric buses instead of diesel ones. This reduces air pollution and makes cities less noisy. In Manchester, 330 electric buses have been introduced to improve public transport while reducing emissions.

Trains are also getting better. Some countries are testing electric and hydrogen-powered trains to replace older diesel ones. In California, a hydrogen-powered train has been launched, and other places are working on similar ideas.

Shared transport is helping too. E-scooters and shared bicycles are now available in many cities. These options help people travel short distances without using cars. In North America, shared transport has reduced carbon emissions by millions of pounds. Carpooling is also becoming more popular, with many people sharing rides to reduce traffic and pollution.

While EVs and cleaner public transport are growing, some problems still need to be solved. In the UK, some people find it hard to charge their EVs because the charging network is not always reliable. Some cities are struggling to keep up with the number of people switching to electric cars. More charging stations and better electricity supply are needed to fix these issues.

More countries are investing in these technologies, and more people are choosing them. As charging stations improve and new transport options are introduced, cleaner travel will become easier for everyone.

Shift to Renewable Energy Sources: Solar, Wind, and Other Clean Energy Technologies

Energy is a part of daily life. It powers homes, businesses, transport, and almost everything people use. For years, coal, oil, and gas have been the main sources of energy, but they release gases that harm the environment. Renewable energy offers a way to generate electricity without adding to pollution. Many countries are moving towards solar, wind, and other cleaner options to replace fossil fuels.

Solar Power

Solar energy is one of the fastest-growing sources of electricity. It works by using panels to capture sunlight and turn it into power. The cost of solar panels has dropped in the last ten years, making it easier for more people and businesses to use them. Governments are providing incentives to install solar panels on homes and buildings to cut down electricity bills and reduce pollution.

Some cities have started using solar farms to generate large amounts of electricity. In India, the Bhadla Solar Park in Rajasthan is one of the biggest in the world, covering over 14,000 acres and producing 2,245 megawatts of electricity. This is enough to power millions of homes. The United States, China, and European countries are also building large solar farms to move away from fossil fuels.

New technology is making solar power even more useful. Scientists are developing solar panels that can generate electricity even on cloudy days or at night by storing extra energy during the day. Some buildings now have solar panels built into windows and rooftops so that they can generate electricity without needing extra space.

Wind Energy

Wind energy is another important clean energy source. Large wind turbines are placed in open areas or offshore to produce electricity from the wind. Countries with strong wind resources, such as the United Kingdom, Denmark, and Germany, have built offshore wind farms in the sea to produce power. These wind farms help supply electricity to millions of homes without using fossil fuels.

The UK has one of the biggest offshore wind farms in the world, Dogger Bank, which will generate enough electricity to power six million homes once fully completed. China and the United States are also expanding wind energy projects.

Some cities are using smaller wind turbines on buildings and open spaces to generate electricity locally. New designs are being developed to make wind turbines quieter and more efficient so they can be used in more places.

Other Clean Energy Technologies

Other forms of clean energy, such as hydropower and geothermal energy, are also being used to replace fossil fuels. Hydropower uses flowing water from rivers and dams to generate electricity. Countries like Norway rely on hydropower for most of their energy needs.

Geothermal energy comes from heat stored beneath the Earth's surface. Countries with natural hot springs, such as Iceland, use this energy to heat homes and generate electricity. The United States, Indonesia, and the Philippines are also increasing their use of geothermal power.

Switching to renewable energy is not only about reducing pollution. It also helps lower electricity costs and makes energy more reliable. Fossil fuels are limited, and their prices keep changing. Renewable energy sources, like the sun and wind, are available every day and do not run out.

Expanding these energy sources will help provide cleaner electricity and create a more stable energy system for the future.

Reducing Plastic Usage: Banning Single-Use Plastics and Promoting Recycling

Plastic pollution is a growing problem worldwide. Each year, about 400 million tonnes of plastic waste are produced, but only 9% is recycled. The rest ends up in landfills, incinerated, or polluting the environment. Single-use plastics, like bags, bottles, and straws, are major contributors to this issue.

Many regions are taking steps to reduce plastic waste by banning single-use plastics. In the United States, more than 500 cities have implemented ordinances to eliminate single-use plastic bags, effectively reducing billions of plastic bags each year. For example, San Francisco was the first city to completely ban plastic bags in 2007, leading to a significant decrease in plastic waste. Similarly, California has introduced legislation to ban the sale of disposable e-cigarettes to combat plastic pollution, aiming to prevent environmental damage from battery acid and other chemicals in single-use vapes.

Internationally, countries like France have enacted laws banning plastic packaging for certain fresh produce items, encouraging consumers to purchase fruits and vegetables without plastic wrapping. This initiative aims to reduce plastic waste and promote sustainable shopping habits.

While reducing the use of single-use plastics is crucial, improving recycling rates is also essential. Currently, global plastic recycling rates are low, with only about 9% of plastic waste being recycled. Efforts are being made to enhance recycling infrastructure and technologies. For instance, startups like Reju in Germany are developing chemical recycling processes that can convert used polyester into high-quality polyester, allowing for continuous recycling without degradation.

Public support for reducing plastic waste is strong. Surveys indicate that more than 80% of people worldwide support a ban on single-use plastics, reflecting a growing awareness of the environmental impact of plastic pollution.

By implementing bans on single-use plastics and investing in advanced recycling technologies, societies can significantly reduce plastic waste and its harmful effects on the environment.

Energy Conservation

Saving energy does not mean making big sacrifices or expensive upgrades. Small changes in daily habits and simple improvements at home can reduce electricity use and lower bills without much effort. Many people do not realise how much energy is wasted because of small things that go unnoticed. Leaving lights on when they are not needed, keeping appliances plugged in even when they are not in use, and using outdated devices that consume more power than necessary all add up over time.

One of the easiest ways to reduce energy use is by choosing appliances that use less electricity. When replacing a fridge, washing machine, or air conditioner, picking a model that is designed to be more efficient can make a difference. These appliances may cost a little more at first, but they save money in the long run because they do not use as much power. Many of them come with labels showing how much electricity they use each year, which helps in making the right choice.

Small home improvements can also help. Closing gaps around windows and doors keeps cool air inside during summer and warm air inside during winter, reducing the need for air conditioning or heating. Using thick curtains can also help maintain indoor temperatures. Replacing old bulbs with LED ones saves electricity, as LED lights last longer and use much less power than traditional bulbs.

Keeping appliances and systems in good condition helps too. Air conditioners and heaters work better when their filters are cleaned regularly. Fridges use less electricity when their coils are dust-free. Even simple things like using a lid while cooking can help food cook faster, reducing the amount of gas or electricity needed.

Many devices continue to use power even when they are switched off. Chargers, televisions, and computers still consume energy as long as they are plugged in. Unplugging them or using a power strip with an off switch can stop this waste.

These small changes do not require much effort but can make a noticeable difference over time.

They help reduce electricity bills and, at the same time, lower the demand for power, which means fewer fossil fuels are burned to generate electricity. Making these adjustments at home is one of the simplest ways to use energy wisely without any inconvenience.

Afforestation and Natural Resource Conservation

Trees are essential for life on Earth. They clean the air, store carbon, protect soil, and support countless species of animals and plants. Large forests act as the planet's lungs, absorbing carbon dioxide and releasing oxygen. With rising global temperatures and extreme weather becoming more common, planting more trees and protecting existing forests has become more important than ever.

Many countries are increasing tree planting efforts to slow down climate change. Afforestation, which means planting trees in areas where there were none before, has been introduced in several regions to capture carbon dioxide from the air. China has planted over 66 billion trees as part of its Great Green Wall project, a massive effort to stop the spread of deserts and restore land. In India, millions of saplings are planted every year as part of environmental campaigns. These projects help reduce pollution, prevent soil erosion, and restore wildlife habitats.

Protecting existing forests is just as important as planting new ones. Trees take decades to grow, but they can be cut down in minutes. The Amazon rainforest, often called the world's lungs, has lost vast areas to deforestation. Illegal logging and land clearing for farming have destroyed large parts of forests, leading to the loss of wildlife and an increase in carbon emissions. Many countries are now enforcing stricter laws to stop deforestation and encourage sustainable land use. Costa Rica is an example of how conservation can work. Once one of the most deforested countries in the world, it has now reversed its damage by restoring more than half of its lost forests through strict protection policies and community-driven reforestation projects.

Forests do more than store carbon. They protect water sources by absorbing rain and keeping rivers clean. When forests disappear, the risk of floods increases, and water supplies dry up faster. Trees also improve soil quality, preventing land from becoming dry and barren. In many parts of Africa, reforestation projects have helped bring back soil fertility, allowing farmers to grow crops in areas that were once considered lost to drought.

Conserving nature goes beyond trees. Wetlands, grasslands, and mangroves also play a big role in maintaining balance in the environment. Mangroves, found along coastlines, act as natural barriers against storms and protect communities from rising sea levels. These ecosystems store even more carbon than rainforests and support marine life. In some areas, damaged mangroves are being restored to prevent further coastal erosion.

People can contribute in different ways. Supporting organisations that focus on tree planting and land conservation is one way to help. Choosing products that come from sustainable sources, such as recycled paper and ethically sourced wood, also reduces pressure on forests. Small actions like planting trees in local communities or reducing waste can have long-term benefits for the environment.

Every tree planted and every natural space protected makes a difference in keeping the planet healthy.

Reducing Consumption

People buy more than they need without realising how much waste it creates. When things are thrown away too soon, it increases demand for new products, leading to more pollution and the use of natural resources. Being more thoughtful about what to buy and how to use things can help reduce waste and stop unnecessary production.

Buying too much causes problems in many ways. Manufacturing goods requires land, water, and energy. The fashion industry alone produces a large amount of carbon emissions because of the fast production of clothes. Many of these clothes are never worn or are thrown away after a short time. Some end up in landfills, while others are shipped to countries that cannot manage the waste. In Ghana, millions of used garments arrive every week, creating mountains of waste that are difficult to dispose of. This is happening with many other products too, from plastic packaging to electronic devices that get replaced too often.

Changing the way we buy things can reduce waste and save resources. Buying fewer but better-quality items means they last longer, reducing the need to replace them often. Second-hand shops and online platforms make it easy to find things that are still in good condition without adding to new production. Some businesses are making products that can be reused or refilled instead of being thrown away after one use. More shops are now offering refill stations where customers can bring their own containers for things like shampoo, cleaning products, and groceries.

Food waste is another issue that can be reduced with simple habits. Planning meals, storing food properly, and using leftovers can help avoid throwing away food unnecessarily. In some places, supermarkets are now selling food that is close to its expiry date at lower prices instead of letting it go to waste.

Being mindful about what to buy and how to use things helps reduce the pressure on natural resources and cuts down on pollution. Small changes in daily choices can make a big difference over time.

What Comes Next

Small changes do not seem like much at first. Buying less, wasting less, using what is already there. It feels like nothing. But over time, these choices shape the way things move.

The café had changed names again. My friend was already there, stirring his coffee, looking at me like he had been waiting to say something.

"You notice how everything is different now?" he said. *"Not all at once, but piece by piece."*

I knew what he meant. A few years ago, plastic bags were everywhere. Now, people carried their own. Shops used to throw out food at the end of the day. Now, they sold it for half the price instead. Buses, cars, bikes, so much of what moved the city ran on electricity instead of fuel.

"It is funny," he said, finishing his coffee. *"One day, things are the way they have always been. Then suddenly, they are not."*

He picked up his bag, ready to leave. *"Next time, we will talk about something new."*

I watched him walk out, thinking about what would come next. What would seem obvious in a few years. What would make people wonder why it had taken so long.

Conclusion

The Earth Was Never Ours to Save

I have lived long enough to see the world change. Some of it was expected. Roads where there were once fields. Taller buildings, fewer open spaces. The kind of change that happens over time, the kind people get used to.

But then there are the changes that do not feel right. The trees that never grew back. The summers that became longer, heavier. The air that feels different, though no one talks about it. The rivers that used to run full but now shrink a little more every year.

I have seen enough to know that this is not the way it was meant to go.

People say we need to save the earth. That if we plant enough trees, cut down on waste, switch to cleaner energy, things might turn around. And maybe they will. Maybe it will be enough. But I have never believed the earth was something to be saved. It has survived worse than us. It has swallowed empires, reshaped itself, wiped away whole histories without a trace. It does not need saving.

We do.

We have built lives that depend on the land, the water, the air. We have taken from the earth as if it owed us something. As if it could keep giving without consequence. And now, when the ground dries up, when the storms last longer, when the heat stays past its welcome, we say we need to fix it.

I do not have answers. I am not here to tell people what to do. I only know what I can do. I can live in a way that does not add to the damage. I can pay attention. I can hold myself accountable for the choices I make. That is my part, small as it may be.

The real question is whether the world will wake up before it is too late. Whether we will accept that the earth does not belong to us, never has. That it is not here for our convenience, not a resource to be managed, not a problem to be solved.

It was meant to sustain us. It was meant to outlive us. If we had listened sooner, maybe it still could.

About the Author

Dr. Manjunatha LR is a civil engineer ,researcher and Management expert with over 32 years of experience in construction, sustainability, and green building materials space. He has always been deeply concerned about the changes in the climate and the impact of human activities on the environment. This book is his way of reflecting on these issues, sharing what he has learned, and exploring solutions that can make a real difference.

He holds a PhD in Management, focusing on the sustainability of green building materials and construction. He also has an MBA and a Bachelor's degree in Engineering. In addition, he is a certified sales trainer from the All India Management Association in New Delhi.

Over the years, he has worked for top corporates like JSW Group, Reliance Industries Limited, Prism Johnson and Zuari cement Limited in India in the building materials industry, Low carbon Cements, Concretes and sustainability sector. He has contributed to professional bodies like the Indian Concrete Institute and the Bureau of Indian Standards, helping to develop sustainability guidelines and promote responsible building practices.

His research on climate action and sustainable construction has reached a wide audience globally, with 75 published research papers and over 95,000 readers with research score of 532 and 114 citations from more then 100 countries across the globe on ResearchGate. He has delivered more than 800 technical lectures and has participated in several national and international conferences.

His work has been recognised with awards such as the CIDC-Vishwakarma Award for Technologist of the Year 2022 and Eminent Civil Engineer of the Year 2022 from the Association of Consulting Civil Engineers, Bangalore.

For Dr. Manjunatha, sustainability is a professional focus, and something he believes in deeply. Through this book, he shares his thoughts on where the world stands today and what can still be done to protect the future.